Wind Energy

Proceedings of the Euromech Colloquium

Joachim Peinke, Peter Schaumann
and Stephan Barth (Eds.)

Wind Energy

Proceedings of the Euromech Colloquium

With 199 Figures and 14 Tables

Prof. Dr. Joachim Peinke
ForWind - Center for Wind Energy Research
Institute of Physics
Carl-von-Ossietzky University Oldenburg
26111 Oldenburg
Germany
peinke@uni-oldenburg.de

Dr. Stephan Barth
ForWind - Center for Wind Energy Research
Institute of Physics
Carl-von-Ossietzky University Oldenburg
26111 Oldenburg
Germany
stephan.barth@uni-oldenburg.de

Prof. Dr.-Ing. Peter Schaumann
ForWind - Center for Wind Energy Research
University of Hannover
Institute for Steel Construction
Appelstrasse 9a
30167 Hannover
Germany
schaumann@stahl.uni-hannover.de

Library of Congress Control Number: 2006932261

ISBN-10 3-540-33865-9 Springer Berlin Heidelberg New York
ISBN-13 978-3-540-33865-9 Springer Berlin Heidelberg New York

This work is subject to copyright. All rights are reserved, whether the whole or part of the material is concerned, specifically the rights of translation, reprinting, reuse of illustrations, recitation, broadcasting, reproduction on microfilm or in any other way, and storage in data banks. Duplication of this publication or parts thereof is permitted only under the provisions of the German Copyright Law of September 9, 1965, in its current version, and permission for use must always be obtained from Springer. Violations are liable to prosecution under the German Copyright Law.

Springer is a part of Springer Science+Business Media.

springer.com

© Springer-Verlag Berlin Heidelberg 2007

The use of general descriptive names, registered names, trademarks, etc. in this publication does not imply, even in the absence of a specific statement, that such names are exempt from the relevant protective laws and regulations and therefore free for general use.

Typesetting by the editors and SPi using Springer LaTeX package
Cover design: Erich Kirchner, Heidelberg

Printed on acid-free paper   SPIN 11534280   89/3100/SPi   5 4 3 2 1 0

Preface

Wind energy is one of the prominent renewable energy sources on earth. During the last decade there has been a tremendous growth, both in size and power of wind energy converters (WECs). The global installed power has increased from 7.5 GW in 1997 to more than 50 GW in 2005 (WWEA – March 2005). At the same time, turbines have grown from kW machines to 5 MW turbines with rotor diameters of more than 100 m. This enormous development and the more recent use in offshore application made high demands on design, construction and operation of WECs. Thus not only a new major industry has been established but also a new interdisciplinary field of research affecting scientists from engineering, physics and meteorology.

In order to tackle the problems and reservations in this interdisciplinary community of wind energy scientists, ForWind, the Center for Wind Energy Research of the Universities of Oldenburg and Hanover, arranged the EUROMECH Colloquium 464b – Wind Energy, which was held from October 4, 7, 2005, at the Carl von Ossietzky University of Oldenburg, Germany. The central aim of this colloquium was to bring together the up to then separate communities of wind energy scientists and those who do fundamental research in mechanics. Wind energy is a challenging task in mechanics and many of future progress will find relevant applications in wind energy conversion.

More than 100 experts coming from 16 countries from all over the world attended the meeting, confirming the need and the concept of this colloquium. The 46 oral and 28 poster presentations were grouped in the following topics:

– Wind climate and wind field
– Gusts, extreme events and turbulence
– Power production and fluctuations
– Rotor aerodynamics
– Wake effects
– Materials, fatigue and structural health monitoring

Phenomenological approaches mainly based on experimental and empirical data as well as advanced fundamental mathematical scientific approaches have

VI   Preface

been presented, spanning the range from reliability investigations to new CFD codes for turbulence models or Levy statistics of wind fluctuations.

During this meeting it became clear, which fundamental scientific tasks will have essential importance for future developments in wind energy:

- A better understanding of the marine atmospheric boundary layer, ranging from mean wind profiles to high resolved influences of turbulence. These questions need further measurements as well as genuine simulations and models. A proper and detailed wind field description is indispensable for correct power and load modeling.
- CFD simulations for wind profiles and rotor aerodynamics with advanced methods (aeroelastic codes) that include experimental details on the dynamic stall phenomenon as well as near and far field rotor wakes.
- A site independent description of wind power production taking into account turbulence induced fluctuations.
- Material loads of different components of a WEC and the fatigue recognition of which due to the high number of lifecycles of such complex machines.
- To establish an advanced numerical hybrid model for a 3D simulation of a WEC, taking into account wind and wave loads as well as all effects of operation in a so-called 'integrated' model.

Many intensive discussions on these and other topics took place between participants from different disciplines during coffee and lunch breaks and also during the social evening events reception of the city at the "ehemalige Exerzierhalle" and the conference dinner on the nightly lake of Bad Zwischenahn.

The positive feedback for the meeting's scientific and social aspects encouraged the scientific committee to decide to have follow-up meetings alternately organized by Duwind, Risø and ForWind. All participants shared the opinion that the scientific interdisciplinary cooperation and international collaboration shall be intensified.

The organizers want to thank the scientific committee members Martin Kühn, Gijs van Kuik, Soeren E. Larsen, Ramgopal Puthli and Daniel Schertzer for helping to organize this conference and establishing this book. Furthermore, we are grateful for the financial support of the Federal Ministry of Education and Research, the City of Oldenburg and the EWE company. Special thanks go to Margret Warns, Elke Seidel, Moses Kärn, Martin Grosser, Frank Böttcher for organizing all technical and administrative concerns.

Contents

List of Contributors XXI

1 Offshore Wind Power Meteorology
Bernhard Lange 1
1.1   Introduction 1
1.2   Offshore Wind Measurements 2
1.3   Offshore Meteorology 2
1.4   Application to Wind Power Utilization 4
1.5   Conclusion 5
References 5

2 Wave Loads on Wind-Power Plants in Deep
and Shallow Water
Lars Bergdahl, Jenny Trumars and Claes Eskilsson 7
2.1   A Concept of Wave Design in Shallow Areas 7
2.2   Deep-Water Wave Data 8
2.3   Wave Transmission into a Shallow Area
       Using a Phase-Averaging Model 8
2.4   Wave Kinematics 10
2.5   Example of Wave Loads 10
2.6   Wave Transmission into a Shallow Area
       Using Boussinesq Models 12
2.7   Conclusions 12
2.8   Acknowledgements 12
References 13

3 Time Domain Comparison of Simulated and Measured
Wind Turbine Loads Using Constrained Wind Fields
Wim Bierbooms and Dick Veldkamp 15
3.1   Introduction 15
3.2   Constrained Stochastic Simulation of Wind Fields . 15

VIII   Contents

3.3   Stochastic Wind Fields which Encompass Measured
       Wind Speed Series.. 16
3.4   Load Calculations Based on Normal and Constrained Wind
       Field Simulations.. 18
3.5   Comparison between Measured Loads and Calculated Ones
       Based on Constrained Wind Fields 19
3.6   Conclusion 20
References 20

4 Mean Wind and Turbulence in the Atmospheric Boundary Layer Above the Surface Layer

S.E. Larsen, S.E. Gryning, N.O. Jensen, H.E. Jørgensen and J. Mann   21
4.1   Atmospheric Boundary Layers at Larger Heights 21
4.2   Data from Høvsøre Test Site 22
4.3   Discussion 24
References 25

5 Wind Speed Profiles above the North Sea

J. Tambke, J.A.T. Bye, B. Lange and J.-O. Wolff 27
5.1   Theory of Inertially Coupled Wind Profiles (ICWP) 27
5.2   Comparison to Observations at Horns Rev and FINO1 29
References 31

6 Fundamental Aspects of Fluid Flow over Complex Terrain for Wind Energy Applications

José Fernández Puga, Manfred Fallen and Fritz Ebert 33
6.1   Introduction 33
6.2   Experimental Setup... 34
6.3   Results.. 35
6.4   Conclusions.. 38
References 38

7 Models for Computer Simulation of Wind Flow over Sparsely Forested Regions

J.C. Lopes da Costa, F.A. Castro and J.M. L.M. Palma 39
7.1   Introduction 39
7.2   Mathematical Models 39
7.3   Results.. 40
7.4   Conclusions.. 42
References 42

8 Power Performance via Nacelle Anemometry on Complex Terrain

Etienne Bibor and Christian Masson 43
8.1   Introduction and Objectives 43
8.2   Experimental Installations 43
8.3   Experimental Analysis 43

Contents IX

8.4	Numerical Analysis	44
8.5	Results and Analysis	44
	8.5.1 Comparaison with the Manufacturer	44
	8.5.2 Influence on the Wind Turbine Control	44
	8.5.3 Influence of the Terrain	45
	8.5.4 Numerical Validation	45
8.6	Conclusion	46
References		47

9 Pollutant Dispersion in Flow Around Bluff-Bodies Arrangement
Elżbieta Moryń-Kucharczyk and Renata Gnatowska 49

9.1	Introduction	49
9.2	Results of Measurements	50
9.3	Conclusions	52
References		52

10 On the Atmospheric Flow Modelling over Complex Relief
Ivo Sládek, Karel Kozel and Zbyněk Jaňour 55

10.1	Mathematical Model	55
	10.1.1 Turbulence Model	56
	10.1.2 Boundary Conditions	56
	10.1.3 Numerical Method	56
10.2	Definition of the Computational Case	57
	10.2.1 Some Numerical Results	58
10.3	Conclusion	59
References		59

11 Comparison of Logarithmic Wind Profiles and Power Law Wind Profiles and their Applicability for Offshore Wind Profiles
Stefan Emeis and Matthias Türk 61

11.1	Wind Profile Laws	61
11.2	Comparison of Profile Laws	61
11.3	Application to Offshore Wind Profiles	62
11.4	Conclusions	64
References		64

12 Turbulence Modelling and Numerical Flow Simulation of Turbulent Flows
Claus Wagner 65

12.1	Summary	65
12.2	Introduction	65
12.3	Governing Equations	66
12.4	Direct Numerical Simulation	67
12.5	Statistical Turbulence Modelling	67

X   Contents

12.6	Subgrid Scale Turbulence Modelling	68
	12.6.1 Eddy Viscosity Models	68
	12.6.2 Scale Similarity Modelling	69
12.7	Conclusion	70
References		70

**13 Gusts in Intermittent Wind Turbulence
and the Dynamics of their Recurrent Times**

François G. Schmitt		73
13.1	Introduction	73
13.2	Scaling and Intermittency of Velocity Fluctuations	74
13.3	Gusts for Fixed Time Increments and Their Recurrent Times	74
13.4	The Dynamics of Inverse Times: Times Needed for Fluctuations Larger than a Fixed Velocity Threshold	78
References		79

**14 Report on the Research Project OWID – Offshore Wind
Design Parameter**

T. Neumann, S. Emeis and C. Illig		81
14.1	Summary	81
14.2	Relevant Standards and Guidelines	81
14.3	Normal Wind Profile	82
14.4	Normal Turbulence Model	82
14.5	Extreme Wind Conditions	84
14.6	Outlook	85
14.7	Acknowledgement	85
References		85

**15 Simulation of Turbulence, Gusts and Wakes for Load
Calculations**

Jakob Mann		87
15.1	Introduction	87
15.2	Simulation over Flat Terrain	87
15.3	Constrained Gaussian Simulation	89
15.4	Wakes	89
	15.4.1 Simulation	89
	15.4.2 Scanning Laser Doppler Wake Measurements	90
References		92

**16 Short Time Prediction of Wind Speeds
from Local Measurements**

Holger Kantz, Detlef Holstein, Mario Ragwitz and Nikolay K. Vitanov		93
16.1	Wind Speed Predictions	93
16.2	Prediction of Wind Gusts	95
References		98

Contents   XI

17 Wind Extremes and Scales: Multifractal Insights and Empirical Evidence
I. Tchiguirinskaia, D. Schertzer, S. Lovejoy and J.M. Veysseire   99
17.1   Atmospheric Dynamics, Cascades and Statistics   99
17.2   Extremes   100
17.3   Discussion and Conclusion   103
References   103

18 Boundary-Layer Influence on Extreme Events in Stratified Flows over Orography
Karine Leroux and Olivier Eiff   105
18.1   Introduction   105
18.2   Experimental Procedure   106
18.3   Basic Flow Pattern   106
18.4   Downstream Slip Condition   107
18.5   Boundary Layer and Wave Field Interaction   108
18.6   Concluding Remarks   109
References   109

19 The Statistical Distribution of Turbulence Driven Velocity Extremes in the Atmospheric Boundary Layer – Cartwright/Longuet-Higgins Revised
G.C. Larsen and K.S. Hansen   111
19.1   Introduction   111
19.2   Model   112
References   114

20 Superposition Model for Atmospheric Turbulence
S. Barth, F. Böttcher and J. Peinke   115
20.1   Introduction   115
20.2   Superposition Model   116
20.3   Conclusions and Outlook   118
References   118

21 Extreme Events Under Low-Frequency Wind Speed Variability and Wind Energy Generation
Alin A. Cârsteanu and Jorge J. Castro   119
21.1   Introduction   119
21.2   Mathematical Background   120
21.3   Results and Conclusions   121
21.4   Acknowledgments   122
References   122

XII   Contents

22 Stochastic Small-Scale Modelling of Turbulent Wind Time Series
Jochen Cleve and Martin Greiner 123
22.1   Introduction 123
22.2   Consistent Modelling of Velocity and Dissipation 123
22.3   Refined Modelling: Stationarity and Skewness 124
22.4   Statistics of the Artificial Velocity Signal 126
References 126

23 Quantitative Estimation of Drift and Diffusion Functions from Time Series Data
David Kleinhans and Rudolf Friedrich 129
23.1   Introduction 129
23.2   Direct Estimation of Drift and Diffusion 130
23.3   Stability of the Limiting Procedure 131
23.4   Finite Length of Time Series 131
23.5   Conclusion 132
References 133

24 Scaling Turbulent Atmospheric Stratification: A Turbulence/Wave Wind Model
S. Lovejoy and D. Schertzer 135
24.1   Introduction 135
24.2   An Extreme Unlocalized (Wave) Extension 136
References 138

25 Wind Farm Power Fluctuations
P. Sørensen, J. Mann, U.S. Paulsen and A. Vesth 139
25.1   Introduction 139
25.2   Test Site 140
25.3   PSDs 141
25.4   Coherence 142
25.5   Conclusion 144
References 145

26 Network Perspective of Wind-Power Production
Sebastian Jost, Mirko Schäfer and Martin Greiner 147
26.1   Introduction 147
26.2   Robustness in a Critical-Infrastructure Network Model 147
26.3   Two Wind-Power Related Model Extensions 151
26.4   Outlook 152
References 152

Contents   XIII

27 Phenomenological Response Theory to Predict Power Output

Alexander Rauh, Edgar Anahua, Stephan Barth and Joachim Peinke .. 153

27.1 Introduction 153
27.2 Power Curve from Measurement Data 154
27.3 Relaxation Model 156
27.4 Discussion and Conclusion 157
References 158

28 Turbulence Correction for Power Curves

K. Kaiser, W. Langreder, H. Hohlen and J. Højstrup 159

28.1 Introduction 159
28.2 Turbulence and Its Impact on Power Curves 160
28.3 Results... 161
28.4 Conclusion 162
References 162

29 Online Modeling of Wind Farm Power for Performance Surveillance and Optimization

J.J. Trujillo, A. Wessel, I. Waldl and B. Lange 163

29.1 Wind Turbine Power Modeling Approach 163
    29.1.1 Wind Farm Model................................. 163
    29.1.2 Online Wind Farm Model 164
29.2 Measurements and Simulation............................... 164
29.3 Results... 165
References 166

30 Uncertainty of Wind Energy Estimation

T. Weidinger, Á. Kiss, A.Z. Gyöngyösi, K. Krassován and B. Papp ... 167

30.1 Introduction 167
30.2 Wind Climate of Hungary 167
30.3 The Uncertainty of the Power Law Wind Profile Estimation 169
30.4 Inter-Annual Variability of Wind Energy 169
30.5 Conclusion 170
References 170

31 Characterisation of the Power Curve for Wind Turbines by Stochastic Modelling

E. Anahua, S. Barth and J. Peinke............................... 173

31.1 Introduction 173
31.2 Simple Relaxation Model 174
31.3 Langevin Method... 175
31.4 Data Analysis.. 175
31.5 Conclusion and Outlook 176
References 177

XIV   Contents

32 Handling Systems Driven by Different Noise Sources: Implications for Power Curve Estimations
F. Böttcher, J. Peinke, D. Kleinhans and R. Friedrich 179
32.1   Power Curve Estimation in a Turbulent Environment 179
      32.1.1   Reconstruction of a Synthetic Power Curve 180
      32.1.2   Additional Noise 182
32.2   Conclusions and Outlook 182
References 182

33 Experimental Researches of Characteristics of Windrotor Models with Vertical Axis of Rotation
Stanislav Dovgy, Vladymyr Kayan and Victor Kochin 183
33.1   Introduction 183
33.2   Experimental Installation and Models 184
33.3   Performance Characteristics of Windrotor Models 184
33.4   Results... 186

34 Methodical Failure Detection in Grid Connected Wind Parks
Detlef Schulz, Kaspar Knorr and Rolf Hanitsch 187
34.1   Problem Description 187
34.2   Doubly-fed Induction Generators 187
34.3   Measurements 188
34.4   Conclusions... 190
References 190

35 Modelling of the Transition Locations on a 30% thick Airfoil with Surface Roughness
Benjamin Hillmer, Yun Sun Chol and Alois Peter Schaffarczyk 191
35.1   Introduction 191
35.2   Measurements 192
35.3   Modelling 192
35.4   Results and Discussion 193
35.5   Conclusions... 195
References 196

36 Helicopter Aerodynamics with Emphasis Placed on Dynamic Stall
Wolfgang Geissler, Markus Raffel, Guido Dietz and Holger Mai 199
36.1   Introduction 199
36.2   The Phenomenon Dynamic Stall 200
36.3   Numerical and Experimental Results for the Typical Helicopter Airfoil OA209 201
36.4   Conclusions... 203
References 204

37 Determination of Angle of Attack (AOA) for Rotating Blades
Wen Zhong Shen, Martin O.L. Hansen and Jens Nørkær Sørensen 205
37.1 Introduction 205
37.2 Determination of Angle of Attack 206
37.3 Numerical Results and Comparisons 207
37.4 Conclusion 209
References 209

38 Unsteady Characteristics of Flow Around an Airfoil at High Angles of Attack and Low Reynolds Numbers
Hui Guo, Hongxing Yang, Yu Zhou and David Wood 211
38.1 Introduction 211
38.2 Test Facility and Setup.................................... 211
38.3 Experimental Results and Discussions...................... 212
38.4 Conclusions... 214
References 214

39 Aerodynamic Multi-Criteria Shape Optimization of VAWT Blade Profile by Viscous Approach
Rémi Bourguet, Guillaume Martinat, Gilles Harran and Marianna Braza 215
39.1 Introduction 215
39.2 Physical Model.. 215
    39.2.1 Templin Method for Efficiency Graphe Computation.... 215
    39.2.2 Flow Simulation.................................... 215
39.3 Blade Profile Optimization 216
    39.3.1 Optimization Method: DOE/RSM 216
    39.3.2 Reaching the Global Optimum 217
39.4 Numerical Results 217
    39.4.1 Validation Results 217
    39.4.2 Optimization Results 217
39.5 Conclusion and Prospects 218
References 218

40 Rotation and Turbulence Effects on a HAWT Blade Airfoil Aerodynamics
Christophe Sicot, Philippe Devinant, Stephane Loyer and Jacques Hureau 221
40.1 Introduction 221
40.2 Experiment 221
40.3 Results and Discussion 222
    40.3.1 Mean Pressure Values Analysis...................... 222
    40.3.2 Instantaneous Pressure Distributions Analysis.......... 224
40.4 Conclusion 225
References 225

XVI   Contents

41 3D Numerical Simulation and Evaluation of the Air Flow Through Wind Turbine Rotors with Focus on the Hub Area
J. Rauch, T. Krämer, B. Heinzelmann, J. Twele and P.U. Thamsen .. 227
41.1   Introduction 227
41.2   Method 228
41.3   Results.. 228
41.4   Perspective 230
References 230

42 Performance of the Risø-B1 Airfoil Family for Wind Turbines
Christian Bak, Mac Gaunaa and Ioannis Antoniou 231
42.1   Introduction 231
42.2   The Wind Tunnel 231
42.3   Results.. 232
42.4   Conclusions.. 233
42.5   Acknowledgements 234
References 234

43 Aerodynamic Behaviour of a New Type of Slow-Running VAWT
J.-L. Menet 235
43.1   Introduction 235
43.2   Description of the Savonius Rotors 236
43.3   Description of the Numerical Model........................ 236
43.4   Results 237
       43.4.1   Optimised Savonius Rotor 237
       43.4.2   The New Rotor 238
43.5   Conclusion 239
References 239

44 Numerical Simulation of Dynamic Stall using Spectral/hp Method
B. Stoevesandt, J. Peinke, A. Shishkin and C. Wagner 241
44.1   Introduction 241
44.2   The Spectral/hp Method 242
44.3   The NekTar Code 243
44.4   First Results... 244
44.5   Outlook 244
References 244

45 Modeling of the Far Wake behind a Wind Turbine
Jens N. Sørensen and Valery L. Okulov.......................... 245
45.1   Extended Joukowski Model 245
45.2   Unsteady Behavior 247

Contents XVII

45.3 Conclusions... 248
References 248

**46 Stability of the Tip Vortices in the Far Wake
behind a Wind Turbine**
Valery L. Okulov and Jens N. Sørensen............................. 249
46.1 Theory: Analysis of the Stability 249
46.2 Application of the Analysis 251
46.3 Conclusions... 251
References 252

47 Modelling Turbulence Intensities Inside Wind Farms
Arne Wessel, Joachim Peinke and Bernhard Lange................. 253
47.1 Description of the Model 253
  47.1.1 Single Wake Model 253
  47.1.2 Superposition of the Wakes 254
47.2 Comparison of the Model with Wake Measurements........... 254
  47.2.1 Vindeby Double and Quintuple Wake 254
47.3 Conclusion 255
References 256

48 Numerical Computations of Wind Turbine Wakes
Stefan Ivanell, Jens N. Sørensen and Dan Henningson.............. 259
48.1 Numerical Method.. 259
48.2 Simulation... 260
References 263

49 Modelling Wind Turbine Wakes with a Porosity Concept
Sandrine Aubrun... 265
49.1 Introduction 265
49.2 Experimental Set-up 265
49.3 Results for Homogeneous Freestream Conditions 266
49.4 Results for Shear Freestream Conditions...................... 267
49.5 Conclusion 269
References 269

**50 Prediction of Wind Turbine Noise Generation
and Propagation based on an Acoustic Analogy**
Dragoş Moroianu and Laszlo Fuchs.................................... 271
50.1 Introduction 271
50.2 Problem Definition 271
50.3 Results... 272
  50.3.1 Flow Computations....................................... 272
  50.3.2 Acoustic Computations 273
  50.3.3 Conclusions 274
References 274

XVIII   Contents

51 Comparing WAsP and Fluent for Highly Complex Terrain Wind Prediction
D. Cabezón, A. Iniesta, E. Ferrer and I. Martí 275
51.1   Introduction 275
51.2   Alaiz Test Site 275
51.3   Description of the Models 276
       51.3.1   Linear Models. WAsP 8.1 (Wind Atlas Analysis
               and Application Program) and WAsP Engineering 2.0 . . 276
       51.3.2   Non Linear Models. Fluent 6.2 276
51.4   Results. 276
       51.4.1   Wind Speed 276
       51.4.2   Turbulence Intensity 279
51.5   Conclusions. 279
References 279

52 Fatigue Assessment of Truss Joints Based on Local Approaches
H. Th. Beier, J. Lange and M. Vormwald 281
52.1   Introduction 281
52.2   Concepts 281
       52.2.1   Fatigue Tests 282
       52.2.2   Crack Initiation with Local Strain Approach 282
       52.2.3   Crack Growth with Linear Elastic Fracture Mechanics . . 283
       52.2.4   Fracture Criterion 284
       52.2.5   Endurance Limit with Local Stress Approach 284
52.3   Examples. 284
       52.3.1   Truss-joint with Pre-cut Gusset Plates (PCGP-joint) . . . 284
       52.3.2   Stiffener of the Great Wind Energy
               Converter GROWIAN 284
52.4   Conclusion 285
References 286

53 Advances in Offshore Wind Technology
Marc Seidel and Jens Gößwein 287
53.1   Introduction 287
53.2   Wind Turbine Technology 287
53.3   Substructure Technology 289
       53.3.1   Design Methodologies 289
       53.3.2   Substructure Concepts 290
53.4   Installation Methods 290
References 291

54 Benefits of Fatigue Assessment with Local Concepts

P. Schaumann and F. Wilke 293

54.1   Introduction 293

54.2   Applied Local Concepts 293

54.3   Comparison of Fatigue Design for a Tripod 294

54.4   Conclusion 296

References 296

55 Extension of Life Time of Welded Fatigue Loaded Structures

Thomas Ummenhofer, Imke Weich and Thomas Nitschke-Pagel 297

55.1   Introduction 297

55.2   Background 297

      55.2.1   Weld Improvement Methods 297

55.3   Experimental Studies 298

      55.3.1   Testing Parameters 298

55.4   Results 298

      55.4.1   Initial State of the Fatigue Test Samples 298

      55.4.2   Results of the Fatigue Tests 299

55.5   Conclusions 300

References 300

56 Damage Detection on Structures of Offshore Wind Turbines using Multiparameter Eigenvalues

Johannes Reetz 301

56.1   Introduction 301

56.2   The Multiparameter Eigenvalue Method 301

56.3   Validation of the Method 303

56.4   Outlook 304

References 304

57 Influence of the Type and Size of Wind Turbines on Anti-Icing Thermal Power Requirements for Blades

L. Battisti, R. Fedrizzi, S. Dal Savio and A. Giovannelli 305

57.1   Introduction 305

57.2   Analysis of the Results 306

57.3   Anti-Icing Power as a Function of the Machine Size 306

57.4   Anti-Icing Power as a Function of the Machine Type 307

57.5   Conclusions 307

References 308

58 High-cycle Fatigue of "Ultra-High Performance Concrete" and "Grouted Joints" for Offshore Wind Energy Turbines

L. Lohaus and S. Anders 309

58.1   Introduction 309

58.2   Ultra-High Performance Concrete 309

XX   Contents

58.3   Ultra-High Performance Concrete in Grouted Joints 310
58.4   Conclusions 311
References 312

59 A Modular Concept for Integrated Modeling of Offshore WEC Applied to Wave-Structure Coupling

Kim Mittendorf, Martin Kohlmeier, Abderrahmane Habbar and Werner Zielke 313

59.1   Introduction 313
59.2   Integrated Modeling 313
       59.2.1   Model Concept 315
       59.2.2   Model Realization 315
59.3   Modeling of Wave Loads on the Support Structure Offshore
       Wind Energy Turbines 316
       59.3.1   Application to the Support Structure of an Offshore
                Wind Turbine 316
59.4   Future Demands 317
References 317

60 Solutions of Details Regarding Fatigue and the Use of High-Strength Steels for Towers of Offshore Wind Energy Converters

J. Bergers, H. Huhn and R. Puthli 319

60.1   Introduction 319
60.2   Fatigue Tests 320
60.3   Finite-Element Analyses 321
References 324

61 On the Influence of Low-Level Jets on Energy Production and Loading of Wind Turbines

N. Cosack, S. Emeis and M. Kühn 325

61.1   Introduction 325
61.2   Data and Methods 325
61.3   Results 326
61.4   Conclusions 327
References 328

62 Reliability of Wind Turbines

Berthold Hahn, Michael Durstewitz and Kurt Rohrig 329

62.1   Introduction 329
62.2   Data Basis 329
62.3   Break Down of Wind Turbines 330
62.4   Malfunctions of Components 331
62.5   Conclusion 332
References 332

List of Contributors

Edgar Anahua
ForWind – Center for
Wind Energy Research
University of Oldenburg
D-26111 Oldenburg
Germany
edgar.anahua@forwind.de

S. Anders
Institute of Building Materials
University of Hannover
Appelstraße 9A, 30161 Hannover
Germany

Ioannis Antoniou
Department of Wind Energy
Risø National Laboratory
P.O. Box 49
DK-4000 Roskilde
Denmark

Sandrine Aubrun
Laboratoire de Mécanique et
d'Energétique, 8 rue Léonard de
Vinci, F-45072 Orléans cedex
France
sandrine.aubrun@univ-orleans.fr

Christian Bak
Department of Wind Energy
Risø National Laboratory
P.O. Box 49
DK-4000 Roskilde
Denmark
christian.bak@risoe.dk

Stephan Barth
ForWind – Center for
Wind Energy Research
University of Oldenburg
D-26111 Oldenburg
Germany
stephan.barth@forwind.de

L. Battisti
DIMS – University of Trento
via Mesiano 77, 38050, Trento
Italy

H. Th. Beier
IFSW, Technische Universität
Darmstadt, Petersenstr. 12
64287 Darmstadt
Germany

J. Bergers
Research Centre for Steel
Timber and Masonry
University of Karlsruhe
Germany

XXII   List of Contributors

Lars Bergdahl
Water Environment Technology
Chalmers, 412 96 Göteborg, Sweden
lars.bergdahl@chalmers.se

Etienne Bibor
Department of Mechanical
Engineering, Ecole de technologie
superieure, 1100 Notre-Dame Ouest
Montreal, Canada
ebibor@hydromega.com

Wim Bierbooms
Delft University of Technology
2629 HS Delft, The Netherlands

Rémi Bourguet
Institut de Mécanique des Fluides
de Toulouse, 6 allée du
Professeur Camille Soula, Toulouse
France
bourguet@imft.fr

F. Böttcher
ForWind – Center for Wind Energy
Research, University of Oldenburg
D-26111 Oldenburg, Germany

Marianna Braza
Institut de Mécanique des
Fluides de Toulouse, 6 allée du
Professeur Camille Soula, Toulouse
France

J.A.T. Bye
The University of Melbourne
Victoria 3010, Australia

D. Cabezón
Department of Wind Energy
National Renewable
Energy Centre (CENER)
C/Ciudad de la Innovacin
31621 Sarriguren, Navarra
Spain

Alin A. Cârsteanu
Department of Mathematics
Cinvestav, Av. IPN 2508, Mexico
D.F. 07360, Mexico

F.A. Castro
CEsA – Research Centre for Wind
Energy and Atmospheric Flows
Faculdade de Engenharia da
Universidade do Porto Rua Roberto
Frias s/n, 4200-465 Porto
Portugal

Jorge J. Castro
Department of Physics, Cinvestav
Av. IPN 2508, Mexico D.F. 07360
Mexico

Yun Sun Chol
Department of Mathematics
and Mechanics, Kim Il Sung
University, Pyongyang
DPR of Korea

Jochen Cleve
Institute of Theoretical Physics
TU Dresden, D-01062 Dresden
Germany
cleve@theory.phy.tu-dresden.de

N. Cosack
Endowed Chair of Wind Energy
Institute of Aircraft Design
University of Stuttgart
Allmandring 5b, 70550 Stuttgart
Germany

S. Dal Savio
DIMS – University of Trento
via Mesiano 77, 38050, Trento
Italy

Philippe Devinant
Laboratoire de Mécanique
et Energétique
Université d'Orléans
8 rue Léonard de Vinci
45072 Orléans, France

Guido Dietz
DLR-Göttingen, Bunsenstr. 10
37073 Göttingen, Germany

Stanislav Dovgy
Institute of Hydromechanics NASU
Kyiv, Ukraine

Michael Durstewitz
Institut für Solare
Energieversorgungstechnik (ISET)
Verein an der Universität
Kassel e.V., 34119 Kassel
Germany

Fritz Ebert
Institute for Mechanical Process
Engineering
University of Kaiserslautern
Erwin-Schrödinger-Strasse 44
67663 Kaiserslautern, Germany
ebert@mv.uni-kl.de

Olivier Eiff
Institut de Mécanique des Fluides
de Toulouse, allée du
Professeur Camille Soula
31400 Toulouse, France
eiff@imft.fr

Stefan Emeis
Institut für Meteorologie und
Klimaforschung, Forschungszentrum
Karlsruhe Kreuzeckbahnstr. 19
Garmisch-Partenkirchen, Germany
stefan.emeis@imk.fzk.de

Claes Eskilsson
Water Environment Technology
Chalmers, 412 96 Göteborg, Sweden
claes.eskilsson@chalmers.se

Manfred Fallen
Institute for Fluid Machinery
and Fluid Mechanics
University of Kaiserslautern
Erwin-Schrödinger-Strasse 44
67663 Kaiserslautern
Germany
fallen@mv.uni-kl.de

R. Fedrizzi
DIMS – University of Trento
via Mesiano 77
38050 Trento, Italy

E. Ferrer
Department of Wind Energy
National Renewable Energy Centre
(CENER)
C/Ciudad de la Innovacin
31621 Sarriguren, Navarra, Spain

Rudolf Friedrich
Westfälische Wilhelms-Universität
Münster
Institut für Theoretische Physik
48149 Münster
Germany

Laszlo Fuchs
Lund University,
Division of Fluid Mechanics
Ole Römersv. 1
P.O. Box 118
22100 Lund
Sweden
laszlo.fuchs@vok.lth.se

Wolfgang Geissler
DLR-Göttingen, Bunsenstr. 10
37073 Göttingen
Germany

A. Giovannelli
University of Rome3, via della Vasca
Navale 79, 00146, Rome

XXIV   List of Contributors

Renata Gnatowska
Institute of Thermal Machinery
Czestochowa University
of Technology
Poland
gnatowska@imc.pcz.czest.pl

Jens Gößwein
REpower Systems AG, Hollesenstr.
15, 24768 Rendsburg, Germany

Martin Greiner
Corporate Technology
Information and Communications
Siemens AG, D-81730 München
Germany
martin.greiner@siemens.com

S.E. Gryning
Department of Wind Energy
Risø, DK-4000, Roskilde
Denmark

Hui Guo
Department of Building Services
Engineering
The Hong Kong Polytechnic
University
Hong Kong, China
and
School of Aeronautical Science
and Engineering
Beijing University of Aeronautics
and Astronautics
Beijing, China

A.Z. Gyöngyösi
Department of Meteorology,
Eötvös University, Pázmány St. 1/A
Budapest, Hungary

Abderrahmane Habbar
ForWind – Center for Wind Energy
Research
Institute of Fluid Mechanics
and Computer Applications in Civil
Engineering University of Hannover
Appelstr. 9A, 30167 Hannover
Germany

Berthold Hahn
Institut für Solare Energiever-
Sorgungstechnik (ISET)
Verein an der Universität Kassel
e.V., 34119 Kassel, Germany

Rolf Hanitsch
Technical University Berlin
Einsteinufer 11
Berlin
Germany
rolf.hanitsch@iee.tu-berlin.de

K.S. Hansen
Technical University of Denmark
DK-2800 Lyngby, Denmark

Martin O.L. Hansen
Department of Mechanical
Engineering
Technical University of Denmark
Building 403
2800 Lyngby
Denmark
molh@mek.dtu.dk

Gilles Harran
Institut de Mécanique des Fluides
de Toulouse, 6 allée du
Professeur Camille Soula, Toulouse
France

B. Heinzelmann
Fluidsystemdynamik
Technische Universität Berlin
Sekr. K2, Straße des 17. Juni 135
10623 Berlin, Germany
bashftfa@mailbox.tu-berlin.de

Dan Henningson
Royal Institute of Technology
Stockholm, Sweden
henning@mech.kth.se

Benjamin Hillmer
Computational Mechanics
Laboratory
University of Applied Sciences
Kiel, Grenzstr. 3
24149 Kiel
Germany
benjamin.hillmer@fh-kiel.de

H. Hohlen
EU Energy Wind Turbines
Seelandstr. 1
23569 Lübeck
Germany

J. Højstrup
Suzlon Energy A/S, Kystvejen 29
8000 Århus, Denmark

Detlef Holstein
Max Planck Institute for
the Physics of Complex Systems
Nöthnitzer Str. 38, 01187 Dresden
Germany

H. Huhn
IMS Ingenieurgesellschaft mbH
Hamburg, Germany

Jacques Hureau
Laboratoire de Mécanique
et Energétique,
Université d'Orléans
8 rue Léonard de Vinci
45072 Orléans, France

C. Illig
DEWI-OCC Offshore and
Certification Centre GmbH
Am Seedeich 9, Cuxhaven
Germany

A. Iniesta
Department of Wind Energy
National Renewable Energy Centre
(CENER)
C/Ciudad de la Innovacin, 31621
Sarriguren, Navarra, Spain

Stefan Ivanell
Royal Institute of Technology
Stockholm, Sweden
Gotland University, Visby
Sweden
stefan.ivanell@hgo.se

Zbyněk Jaňour
Institute of Thermomechanics
Czech Academy of Sciences
Dolejškova 5, ZIP 182 00, Prague
Czech Republic
janour@it.cas.cz

N.O. Jensen
Department of Wind Energy
Risø DK-4000, Roskilde
Denmark

H.E. Jørgensen
Department of Wind Energy, Risø
DK-4000, Roskilde, Denmark

Sebastian Jost
Corporate Technology
Information and Communications
Siemens AG
D-81730 München, Germany
josts@cip.ifi.lmu.de

K. Kaiser
Ingenieurbüro, Gr. Burgstr. 27
23552 Lübeck, Germany

Holger Kantz
Max Planck Institute for the Physics
of Complex Systems,
Nöthnitzer Str. 38, 01187 Dresden
Germany
kantz@mpipks-dresden.mpg.de

Vladymyr Kayan
Institute of Hydromechanics NASU
Kyiv, Ukraine
kayan@ua.fm

XXVI   List of Contributors

A. Kiss
Department of Atomic Physics
Eötvös University, Pázmány
St. 1/A, Budapest, Hungary

David Kleinhans
Westfälische Wilhelms-Universität
Münster, Institut für Theoretische
Physik
48149 Münster
Germany

Kaspar Knorr
Technical University Berlin
Einsteinufer 11, Berlin, Germany

Victor Kochin
Institute of Hydromechanics NASU
Kyiv, Ukraine

Martin Kohlmeier
ForWind – Center for Wind Energy
Research
Institute of Fluid Mechanics and
Computer Applications in Civil
Engineering University of Hannover
Appelstr. 9A, 30167 Hannover
Germany

Karel Kozel
Czech Technical University
in Prague, U12101, Karlovo
náměstí 13, ZIP 121 35
Czech Republic
kozelk@fsik.cvut.cz

T. Krämer
Fluidsystemdynamik, Technische
Universität Berlin, Sekr. K2
Straße des 17. Juni 135
10623 Berlin, Germany

K. Krassován
Department of Atomic Physics
Eötvös University, Pázmány St.
1/A, Budapest, Hungary

M. Kühn
Endowed Chair of Wind Energy
Institute of Aircraft Design
University of Stuttgart
Allmandring 5b, 70550 Stuttgart
Germany

Bernhard Lange
ISET e.V., Königstor 59
34119 Kassel, Germany
blange@iset.uni-kassel.de

J. Lange
IFSW, Technische Universität
Darmstadt, Petersenstr. 12
64287 Darmstadt, Germany

W. Langreder
Wind Solutions,
Engelsgrube 25 23552 Lübeck
Germany

G.C. Larsen
Department of Wind Energy
Risø National Laboratories
DK-4000 Roskilde, Denmark

S.E. Larsen
Department of Wind Energy, Risø
DK-4000, Roskilde, Denmark

Karine Leroux
Centre National de Recherches
Météorologiques de Toulouse
Météo-France, 42 av. G. Coriolis
31057 Toulouse Cedex, France
and
Institut de Mécanique des Fluides
de Toulouse, allée du
Professeur Camille Soula
31400 Toulouse, France
karine.leroux@cnrm.meteo.fr

L. Lohaus
ForWind – Center for Wind Energy
Research
Institute of Building Materials
University of Hannover
Appelstraße 9A, 30161 Hannover
Germany

J.C. Lopes da Costa
CEsA – Research Centre for Wind
Energy and Atmospheric Flows
Faculdade de Engenharia da
Universidade do Porto Rua
Roberto Frias s/n
4200-465 Porto, Portugal

S. Lovejoy
Physics, McGill University, 3600
University St., Montreal, Que.
Canada

Stephane Loyer
Laboratoire de Mécanique
et Energétique
Université d'Orléans
8 rue Léonard de Vinci 45072
Orléans, France

Mac Gaunaa
Department of Wind Energy
Risø National Laboratory
P.O. Box 49
DK-4000 Roskilde, Denmark

Holger Mai
DLR-Göttingen, Bunsenstr. 10
37073 Göttingen, Germany

Jakob Mann
Wind Energy Department
Risø National Laboratory, VEA-118
P.O. Box 49
DK-4000 Roskilde
Denmark
jakob.mann@risoe.dk

I. Martí
Department of Wind Energy
National Renewable Energy Centre
(CENER)
C/Ciudad de la Innovacin
31621 Sarriguren, Navarra, Spain

Guillaume Martinat
Institut de Mécanique des Fluides
de Toulouse, 6 allée du
Professeur Camille Soula, Toulouse
France
martinat@imft.fr

Christian Masson
Department of Mechanical
Engineering
Ecole de technologie superieure
1100 Notre-Dame Ouest
Montreal, Canada
christian.masson@etsmtl.ca

J.-L. Menet
Laboratoire de Mécanique
et d'Énergétique – Valenciennes
University Le Mont Houy 59313
Valenciennes Cedex 9, France

Kim Mittendorf
ForWind – Center for Wind Energy
Research
Institute of Fluid Mechanics
and Computer Applications
in Civil Engineering
University of Hannover
Appelstr. 9A, 30167 Hannover
Germany
mdorf@hydromech.uni-hannover.de

Dragoş Moroianu
Lund University
Division of Fluid Mechanics
Ole Römersv. 1
P.O. Box 118
22100 Lund
Sweden
dragos.moroianu@vok.lth.se

XXVIII   List of Contributors

Elżbieta Moryń-Kucharczyk
Institute of Thermal Machinery
Czestochowa University of
Technology
Poland
moryn@imc.pcz.czest.pl

T. Neumann
DEWI German Wind Energy
Institute
Ebertstr. 96 Wilhelmshaven
Germany

Thomas Nitschke-Pagel
Institut für Füge- und
Schweißtechnik, Technische
Universität Braunschweig, Langer
Kamp 8, Braunschweig, Germany
t.pagel@tu-bs.de

Valery L. Okulov
Department of Mechanical
Engineering
Technical University of Denmark
DK-2800 Lyngby
Denmark
and
Institute of Thermophysics
SB RAS Lavrentyev Ave. 1
Novosibirsk 630090
Russia

J.M.L.M. Palma
CEsA – Research Centre for Wind
Energy and Atmospheric Flows
Faculdade de Engenharia da
Universidade do Porto Rua Roberto
Frias s/n, 4200-465 Porto
Portugal
jpalma@fe.up.pt

B. Papp
Department of Atomic Physics
Eötvös University
Pázmány St. 1/A
Budapest, Hungary

U.S. Paulsen
Risø National Laboratory, VEA-118
PO Box 49, DK-4000 Roskilde
Denmark

Joachim Peinke
ForWind – Center for Wind Energy
Research
University of Oldenburg
D-26111 Oldenburg
Germany

José Fernández Puga
Institute for Mechanical Process
Engineering
University of Kaiserslautern
Erwin-Schrödinger-Strasse 44
67663 Kaiserslautern
Germany
fernandez@mv.uni-kl.de

R. Puthli
Research Centre for Steel
Timber and Masonry
University of Karlsruhe
Germany

Markus Raffel
DLR-Göttingen, Bunsenstr. 10
37073 Göttingen, Germany

Mario Ragwitz
Fraunhofer Institute for Systems
and Innovation Research
Breslauer Str. 48, 76139
Karlsruhe
Germany

J. Rauch
Fluidsystemdynamik, Technische
Universität Berlin, Sekr. K2
Straße des 17. Juni 135
10623 Berlin, Germany
bashftfa@mailbox.tu-berlin.de

Alexander Rauh
Institute of Physics
University of Oldenburg
D-26111 Oldenburg
Germany
alexander.rauh@uni-oldenburg.de

Johannes Reetz
ForWind – Center for Wind Energy
Research
Institute for Structural Analysis
University of Hannover
Germany

Kurt Rohrig
Institut für Solare
Energieversorgungstechnik (ISET)
Verein an der
Universität Kassel e.V., 34119
Kassel, Germany

Mirko Schäfer
Corporate Technology
Information and Communications
Siemens AG
D-81730 München, Germany
mirko.schaefer@theo.physik.
unigiessen.de

Alois Peter Schaffarczyk
Computational Mechanics
Laboratory
University of Applied Sciences, Kiel
Grenzstr. 3, 24149 Kiel
Germany

P. Schaumann
ForWind – Center for Wind Energy
Research
Institute for Steel Construction
University of Hannover
Appelstraße 9A, 30167 Hannover
Germany

D. Schertzer
CEREVE, ENPC, 6-8, ave. Blaise
Pascal, Cité Descartes, 77455
Marne-la-Vallée Cedex, France

Météo-France, Paris, 1 Quai
Branly, 75007 Paris, France

François G. Schmitt
CNRS, FRE ELICO 2816, Station
Marine de Wimereux, Université de
Lille 1, 28 av. Foch
62930 Wimereux
France
francois.schmitt@univ-lille1.fr

Detlef Schulz
University of Applied Sciences
Bremerhaven/Competence Center
Wind Energy
An der Karlstadt 8, Bremerhaven
Germany
dschulz@hs-bremerhaven.de

Marc Seidel
REpower Systems AG
Hollesenstr 15, 24768 Rendsburg
Germany
m.seidel@repower.de

Wen Zhong Shen
Department of Mechanical
Engineering
Technical University of Denmark
Building 403
2800 Lyngby, Denmark
shen@mek.dtu.dk

A. Shishkin
DLR Göttingen, Robert-Bunsenstr.
10, D-37073 Göttingen, Germany

Christophe Sicot
Laboratoire de Mécanique
et Energétique
Université d'Orléans
8 rue Léonard de Vinci 45072
Orléans, France
christophe.sicot@univ-orleans.fr

XXX List of Contributors

Ivo Sládek
Czech Technical University
in Prague U12101, Karlovo
náměstí 13, ZIP 121 35
Czech Republic
sladek@marian.fsik.cvut.cz

Jens Nørkær Sørensen
Department of Mechanical
Engineering
Technical University of Denmark
Building 403, DK-2800
Lyngby, Denmark
jns@mek.dtu.dk

P. Sørensen
Risø National Laboratory, VEA-118
P.O. Box 49
DK-4000 Roskilde, Denmark

B. Stoevesandt
ForWind, Marie-Curie-Str. 2
26129 Oldenburg, Germany
bernhard.stoevesandt@forwind.de

J. Tambke
ForWind – Center for Wind Energy
Research
Carl von Ossietzky University
Oldenburg, 26111 Oldenburg
Germany

I. Tchiguirinskaia
CEREVE, ENPC, 6-8, av. Blaise
Pascal, Cité Descartes, 77455
Marne-la-Vallée cedex, France

P.U. Thamsen
Fluidsystemdynamik
Technische Universität Berlin
Sekr. K2, Straße des 17.
Juni 135, 10623 Berlin, Germany

J.J. Trujillo
ForWind – Oldenburg University
now at SWE Stuttgart University
Germany
juanjose.trujillo@forwind.de
juan-jose.trujillo@ifb.
unistuttgart.de

Jenny Trumars
Water Environment Technology
Chalmers, 412 96 Göteborg
Sweden
jenny.trumars@chalmers.se

Matthias Türk
Institut für Meteorologie und
Klimaforschung, Forschungszentrum
Karlsruhe Kreuzeckbahnstr. 19
Garmisch-Partenkirchen
Germany

J. Twele
Fluidsystemdynamik
Technische Universität Berlin
Sekr. K2
Straße des 17. Juni 135
10623 Berlin
Germany

Thomas Ummenhofer
Institut für Bauwerkserhaltung
und Tragwerk
Technische Universität Braunschweig
Pockelsstr.3, Braunschweig
Germany
t.ummenhofer@tu-bs.de

Dick Veldkamp
Vestas Wind Systems A/S Alsvej 21
8900 Randers, Denmark

A. Vesth
Risø National Laboratory, VEA-118
P.O. Box 49, DK-4000 Roskilde
Denmark

J.M. Veysseire
Direction de la Climatologie
Météo-France, 42 Av. G. Coriolis
31057 Toulouse Cedex
France

Nikolay K. Vitanov
Institute of Mechanics
Bulgarian Academy of Sciences
Akad. G. Bonchev Str.
Block 4, 1113, Sofia, Bulgaria

M. Vormwald
IFSW, Technische Universität
Darmstadt, Petersenstr. 12
64287 Darmstadt, Germany
vormwald@wm.tu-darmstadt.de

Claus Wagner
German Aerospace Center
Institute for Aerodynamics
and Flow Technology
Bunsenstr. 10
D-37073 Göttingen, Germany
claus.wagner@dlr.de

I. Waldl
Overspeed GmbH & Co. KG
Oldenburg, Germany
igor@overspeed.de

Imke Weich
Institut für Bauwerkserhaltung
und Tragwerk
Technische Universität Braunschweig
Pockelsstr.3, Braunschweig
Germany
i.weich@tu-bs.de

T. Weidinger
Department of Meteorology
Eötvös University, Pázmány st.
1/A, Budapest, Hungary
zeno@nimbus.elte.hu

Arne Wessel
ForWind
University Oldenburg
Marie-Curie-Str. 1, Oldenburg
Germany
arne.wessel@forwind.de

F. Wilke
ForWind – Center for Wind Energy
Research
Institute for Steel Construction
University of Hannover
Appelstraße 9A, 30167 Hannover
Germany

J.-O. Wolff
Carl von Ossietzky University
Oldenburg
26111 Oldenburg
Germany

David Wood
School of Engineering
University of Newcastle
Callaghan, Australia

Hongxing Yang
Department of Building Services
Engineering
The Hong Kong Polytechnic
University
Hong Kong
China

Yu Zhou
Department of Mechanical
Engineering, The Hong Kong
Polytechnic University
Hong Kong

Werner Zielke
ForWind – Center for Wind Energy
Research
Institute of Fluid Mechanics
and Computer Applications
in Civil Engineering
University of Hannover
Appelstr. 9A, 30167 Hannover
Germany

1

Offshore Wind Power Meteorology

Bernhard Lange

Summary. Wind farms built at offshore locations are likely to become an important part of the electricity supply of the future. For an efficient development of this energy source, in depth knowledge about the wind conditions at such locations is therefore crucial. Offshore wind power meteorology aims to provide this knowledge. This paper describes its scope and argues why it is needed for the efficient development of offshore wind power.

1.1 Introduction

Wind power utilization for electricity production has a huge resource and has proven itself to be capable of producing a substantial share of the electricity consumption. It is growing rapidly and can be expected to contribute substantially to our energy need in the future (GWEC, 2005). The 'fuel' of this electricity production is the wind. The wind is, on the other hand, also the most important constraint for turbine design, as it creates the loads the turbines have to withstand.

Therefore, accurate knowledge about the wind is needed for planning, design and operation of wind turbines. Some tasks where specific meteorological knowledge is essential are wind turbine design, resource assessment, wind power forecasting, etc. Wind power meteorology has therefore established itself as an important topic in applied meteorology (Petersen et al., 1998). For wind power utilization on land, substantial knowledge and experience has been gained in the last decades, based on the detailed meteorological and climatological knowledge available. Offshore, the meteorological knowledge is less developed since there has been little need to know the wind at heights of wind turbines over coastal waters and any measurements at offshore locations are difficult and extremely expensive.

The aim of this paper is to describe the scope of offshore wind power meteorology and to argue why this topic should be given more attention both from the meteorological point of view and from the wind power application point of view. The paper is structured in three main sections: First some particular problems of offshore measurements are discussed in Sect. 1.2. This is followed by a section giving examples for meteorological effects specific for offshore conditions. Their importance for wind power application is shown in Sect. 1.4, followed by the conclusion.

1.2 Offshore Wind Measurements

In recent years, measurements with the aim to determine the wind conditions for offshore wind power utilization have been erected at a number of locations (Barthelmie et al., 2004). Offshore wind measurements are a challenging task, not only since an offshore foundation and support structure for the mast are needed, but also because of the challenges to provide an autonomous power supply and data transfer, the difficulties of maintenance and repair in an offshore environment, etc. These difficulties lead to high costs of offshore measurements and often lower data availability compared to locations on land. Additionally, the flow distortion of the self supporting mast usually requires a correction of the measured wind speeds for wind profile measurements (Lange, 2004).

Two measurements, from which results are shown in this paper, are the Rødsand field measurement in the Danish Balitc Sea and the FINO 1 measurement in the German Bight. The FINO 1 measurement platform (Rakebrandt-Gräßner and Neumann, 2003) is located 45 km north of the island Borkum in the North Sea (see Fig. 1.1). The height of the measurement mast is 100 m. The field measurement program Rødsand (Lange et al., 2001) is situated about 11 km south of the island Lolland in Denmark (see Fig. 1.1) and includes a 50 m high meteorological mast.

1.3 Offshore Meteorology

There are fundamental differences between the wind conditions over land and offshore due to the influence of the surface on the flow. The most obvious one is the roughness of the sea, which is very low, but also changes due to the changing wave field (Lange et al., 2004b). The momentum transfer between

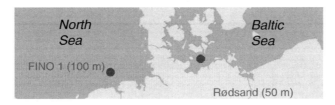

Fig. 1.1. The measurement sites Rødsand in Denmark and FINO 1 in Germany

wind and water, governed by the sea surface roughness, therefore depends on the wave field (see Fig. 1.2).

Stability effects due to the different thermal properties of water compared to land have been shown to be very important (Barthelmie, 1999), (Lange et al., 2004). Both the surface roughness and the surface temperature change abruptly at the coastline, which leads to important transition effects for wind blowing from land to the sea. Additionally, other effects like currents and tides influence the wind speed over water (Barthelmie, 2001).

The dedicated meteorological measurements made in connection with planned offshore wind power development helped to improve the knowledge about the wind conditions relevant for offshore wind farm installations. One example is the vertical wind speed profile over coastal waters.

The wind speed profile is commonly described by a logarithmic profile, modified by Monin–Obukhov similarity theory for thermal stability. In Fig. 1.3 the prediction of Monin–Obukhov theory for the ratio of wind speeds at 50 m

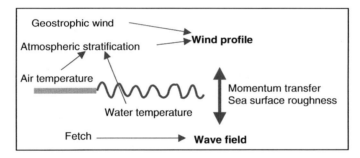

Fig. 1.2. Sketch of influences on the wind field over coastal waters

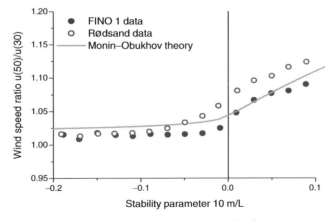

Fig. 1.3. Comparison of measured (Rødsand and FINO 1) and theoretical (Monin–Obukhov theory) dependence of the wind speed ratio at the heights 50 and 30 m on atmospheric stability

and 30 m height versus stability is shown together with measured results from the two sites Rødsand and FINO 1 (Lange, 2004). It can be seen that the Rødsand data show a larger wind speed ratio for near neutral and stable conditions than expected from theory.

A qualitative explanation of this result based on (Csanady, 1974) has been developed (Lange et al., 2004): Rødsand is surrounded by land in all directions with a distance to the coast of 10 to 100 km. When warm air is advected from land over a colder sea, an internal boundary layer with stable stratification develops at the coastline. The heat flow through the stable layer is small, and the air close to the water is cooled continuously from the sea surface. It will eventually take the temperature of the sea and become a well-mixed layer with near-neutral stratification. Higher up an inversion develops with strongly stable stratification. In such a situation with strong height inhomogeneity of atmospheric heat flux, Monin–Obukhov theory must fail. At the FINO 1 site, the coastline is much further away for almost all wind directions and this flow situation does not develop.

1.4 Application to Wind Power Utilization

For planning and operation of offshore wind farms, it is important to take into account the specific conditions at offshore locations. As shown above, the vertical wind speed profile can be modified significantly in the coastal zone. A simple correction method has been proposed to evaluate the magnitude of the effect for wind power applications (Lange et al., 2004a). The effect of this correction on the profile can be seen in Fig. 1.4, where different theoretical wind profiles are compared.

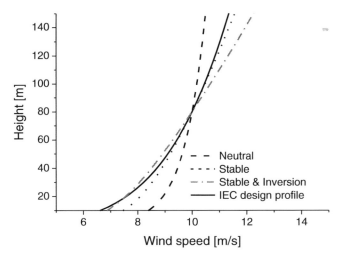

Fig. 1.4. Comparison of different theoretical wind speed profiles

The logarithmic profile expected for neutral stratification, a Monin–Obukhov profile for stable stratification (L=200 m) and a profile additionally taking into account the effect of an inversion (h=200 m) (Lange et al., 2004a). Clearly, the wind speed gradient with height increases when the inversion is included. The gradient is then larger than the gradient of the power law profile used in the IEC guidelines (IEC-61400-1, 1998) for wind turbine design, which do not take atmospheric stability into account. This means that the fatigue loads on e.g. the blades will in these situations be larger than anticipated in the design guidelines. Over land stability is always near neutral at high wind speeds due to the low surface roughness. Over water, on the other hand, stable stratification also occurs at higher wind speeds. Therefore, atmospheric stability might have to be included in the description of the wind shear.

1.5 Conclusion

With the example of the vertical wind speed profile offshore it was shown that specific meteorological conditions exist at the potential locations of offshore wind farms, i.e. over coastal waters in heights of 20 to 200 m. Since the interest in the wind conditions at these locations is new, the specific meteorological knowledge still has to be improved. The behaviour of the atmospheric flow over the sea differs from what is seen over land due to the different properties of the water surface. The findings still have to be investigated further, but it is clear that specifically offshore wind conditions can have important effects on wind power utilization, e.g. for turbine design and wind resource calculation. This leads to the conclusion that offshore wind power meteorology is an important research field, which is needed for the efficient development of offshore wind power and which has the potential to produce new meteorological knowledge about the atmospheric flow over the sea.

References

1. Barthelmie RJ (1999) The effects of atmospheric stability on coastal wind climates. Meteorological Applications 6(1): 39–47
2. Barthelmie RJ (2001) Evaluating the impact of wind induced roughness change and tidal range on extrapolation of offshore vertical wind speed profiles. Wind Energy 4: 99–105
3. Barthelmie RJ, Hansen O, Enevoldsen K, Motta M, Højstrup J, Frandsen S, Pryor S, Larsen S, Sanderhoff P (2004) Ten years of measurements of offshore wind farms – What have we learnt and where are the uncertainties? In: Proceedings of the EWEA Special Topic Conference, Delft, The Netherlands
4. Csanady GT (1974) Equilibrium theory of the planetary boundary layer with an inversion lid, Bound-Layer Meteor. 6: 63–79
5. GWEC Wind Force 12 (2005) A blueprint to achieve 12% of the world's electricity from wind power by 2020. (available from www.ewea.org)

6   B. Lange

6. IEC-61400-1 (1998) Wind turbine generator systems part 1: Safety requirements. International Electrotechnical Commission, Geneva, Swiss
7. Lange B (2004). Comparison of wind conditions of offshore wind farm sites in the Baltic and North Sea. In: Proceedings of the German Wind Energy Conference DEWEK 2004, Wilhelmshaven, Germany
8. Lange B, Barthelmie RJ, Højstrup J (2001) Description of the Rødsand field measurement. Risø-R-1268, Risø National Laboratory, Roskilde, Denmark
9. Lange B, Larsen S, Højstrup J, Barthelmie RJ (2004) The influence of thermal effects on the wind speed profile of the coastal marine boundary layer. Bound-Layer Meteor. 112: 587–617
10. Lange B, Larsen S, Højstrup J, Barthelmie RJ (2004a) Importance of thermal effects and sea surface roughness for offshore wind resource assessment. Journal of Wind Engineering and Industrial Aerodynamics 92 (11): 959–998
11. Lange B, Johnson HK, Larsen S, Højstrup J, Kofoed-Hansen H, Yelland MJ (2004b) On detection of a wave age dependency for the sea surface roughness. Journal of Physical Oceanography 34: 1441–1458
12. Petersen EL, Mortensen NG, Landberg L, Højstrup J, Frank HP (1998) Wind power meteorology. Part I: climate and turbulence. Wind Energy 1(1): 2–22, Part II: siting and models. Wind Energy 1(2): 55–72
13. Rakebrandt-Gräßner P, Neumann T (2003) The German Research Platform in the North Sea. In: Proceedings of the OWEMES 2003, Naples, Italy

2

Wave Loads on Wind-Power Plants in Deep and Shallow Water

Lars Bergdahl, Jenny Trumars and Claes Eskilsson

Summary. A concept for describing design waves for a near-shore site of a wind-power plant and ultimately the wave loads is to transform the off-coast wave spectrum to the target site by a model for wave transformation. At the site second order, irregular, non-linear, shallow-water waves are subsequently realized in the time domain. Alternatively a Boussinesq model is used. Finally in the examples here Morison's equation is used for the wave load and overturning moment.

2.1 A Concept of Wave Design in Shallow Areas

Usually there is little knowledge of long-term wave conditions at prospective sites for wind-power plants, while the deep-water or open sea conditions may be more known and geographically less varying. Then a concept for assessing design waves for the site and ultimately wave loads would be to transform the off-coast waves to the target near-coast site or shallow offshore shoal by some model for the wave transformation. Such models can be divided into two general classes: phase-resolving models, which model the progression of the physical "wave train", predicting both amplitudes and phases of individual waves, and phase-averaging models, which model the progression of average quantities such as the wave spectrum or its integral properties (e.g. H_s, T_z). Here examples of using phase average models (WAM and SWAN) and a phase resolving model (Boussinesq) will be demonstrated. Using e.g. the phase-averaging model SWAN for the transformation to the site, it is subsequently necessary to make a time realization of the transformed wave spectrum into the time domain as the loads on a slender structure is due to non-linear drag forces, the instantaneous elevation of the water surface and – for high waves – the skewness of the elevation. For a phase resolving method the transformed wave is already in the time domain and can thus be used "directly" in the load modelling.

$$M = \frac{1}{2}\rho C_{\mathrm{D}} D \int_{-h}^{\eta} u \left| u \right| (z + h) \mathrm{d}z + \rho \frac{\pi D^2}{4} C_{\mathrm{M}} \int_{-h}^{\eta} u_t (z + h) \mathrm{d}z \qquad (2.1)$$

In the examples here Morison's equation is used for the wave load and overturning moment, (2.1), where u is the horizontal water velocity, z the vertical coordinate and h the water depth. The aim of the load calculation can be to assess extreme loads or fatigue. In both cases non-linear wave properties may be important, but for the extreme loads sometimes a monochromatic design wave may be sufficient. The deep-water waves are usually considered linear. Then a Gaussian-distributed stochastic process symmetric around the mean water elevation can model the time and space varying wavy surface. For steep waves this is not correct. The wave crests are higher and sharper while the wave troughs are shallower and flatter than in the Gaussian model. In shallow areas the non-linearities are further amplified by the influence of the bottom.

2.2 Deep-Water Wave Data

The deep-water wave climate is not sufficiently well known. Wave measurements were initiated in Swedish water at a few places in the Swedish wave-energy programme in the 1970s but have not been much evaluated. A more viable possibility for the Baltic is to use wave-data from the WAM4 [1] model erected for the Baltic Sea and run at ICM in Warsaw [2]. The Baltic WAM4 model is applied to a quadrilateral grid with $0.15°$ (ca 16.7 km). To validate the model it was run for periods during which, also waves were measured ca 8 km off the Polish coast with directional wave rider buoys. The significant wave height was chosen for comparison. For onshore winds $\pm 100°$ the correlation coefficient between WAM4 waves and measured waves was above 0.8 [3].

2.3 Wave Transmission into a Shallow Area Using a Phase-Averaging Model

In a Swedish investigation on wave loading on the Bockstigen wind-power plant [4] SWAN [5] was used to transfer deep-water waves closer to shore. The position and bottom topography for Bockstigen is shown in Fig. 2.1. As an example using a sea state defined by a JONSWAP spectrum $H_s = 4.5$ m, $T_p = 6.7$ s, $\cos^2$ spreading and wind velocity 20 m/s from southwest as input to the model the resulting output inshore at Bockstigen was $H_s = 2.7$ m, $T_p = 8.9$ s, so energy has been dissipated but also shifted in the frequency domain. Especially the inshore spectrum exhibits a secondary hump around the double peak frequency, which is important and typical for shoaling waves. The waves of this hump may be a mixture of short first-order waves and bound waves propagating with the same celerity as the primary peak waves. The pressure and particle velocity in the bound waves attenuate slower with depth than corresponding linear waves.

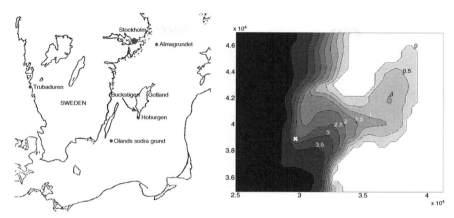

Fig. 2.1. The Island of Gotland with Bockstigen and the bottom topography there

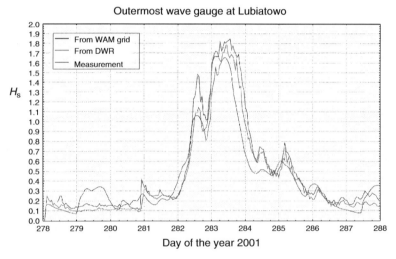

Fig. 2.2. A comparison between measured significant wave height for the wave staff closest to land at Lubiatowo and wave heights modelled from wave rider (DWR) and WAM data 5 km offshore

SWAN has been validated for a surf zone with four alongshore bars on the Polish coast [6]. Two different inputs to the model were given: modelled WAM data and measured data from a directional wave rider buoy 5 km off the coast. Modelled SWAN data were compared to simultaneous wave measurements taken with three wave staffs (capacitance gauges) around 200, 400 and 550 m off the coast. Comparison of modelled and measured data is shown in Fig. 2.2. The agreement seems to be good enough for engineering purposes.

2.4 Wave Kinematics

The wave kinematics can be realized to second order for weakly non-linear waves in deep and shallow water. Equation (2.2) and (2.3) give the first and second order contributions. $F(n)$ are the first-order component amplitudes, ω_n the angular frequencies and k_n the wave numbers. $H(\omega_n, \omega_m)$ is a quadratic transfer function. For details see [7].

$$_1\eta(x,t) = \sum_{m=0}^{N} F(m) \exp i(\omega_m t - k_m x) \qquad (2.2)$$

$$_2\eta(x,t) = \sum_{n=0}^{N} \sum_{m=0}^{N} \frac{F(n)F(m)}{4g} H(\omega_n, \omega_m) \exp i\left[(\omega_n + \omega_m)t - (k_n + k_m)x\right] \qquad (2.3)$$

The equations are valid to the mean water elevation and have to be extrapolated to the instantaneous water surface.

2.5 Example of Wave Loads

In Figs. 2.3 and 2.4 comparisons of linear (1st order) and non-linear (1st+2nd order) realizations of forces and moments are shown [4]. For these high waves

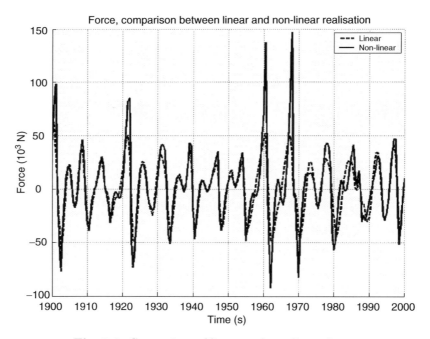

Fig. 2.3. Comparison of linear and non-linear force

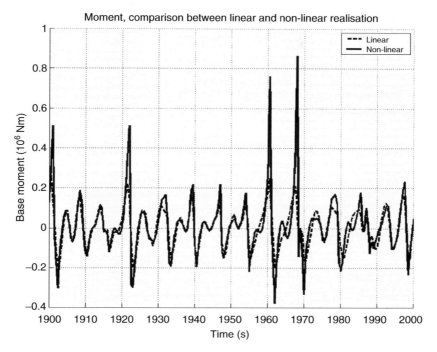

Fig. 2.4. Comparison of linear and non-linear overturning moment

Table 2.1. Accumulated fatigue damage for the whole lifetime of a structure using a linear S–N curve with $m = 3$[1] NL/L is the ratio between non-linear and linear damage

	1	2	3	4	5	6	Mean
Linear	0.55	0.50	0.57	0.49	0.56	0.55	0.54
Non-linear	0.71	0.63	0.63	0.58	0.72	0.67	0.66
No wave	0.08	0.09	0.07	0.07	0.09	0.09	0.08
NL/L	1.29	1.25	1.11	1.19	1.28	1.22	1.22

the difference is large. In Table 2.1 simulated fatigue damages – stress due to overturning moment – to a wind turbine in 20 m water depth are listed for linear waves, non-linear waves and wind only [7]. The six sets of simulations give as a mean 20% larger damage for non-linear waves.

---

[1] The stress amplitude, S, as a function of No of load cycles to failure, N: log(N) = log(K) − m log(S), K and m are empirical material parameters.

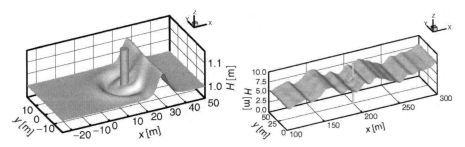

Fig. 2.5. A solitary wave and irregular waves passing a cylinder [8]

2.6 Wave Transmission into a Shallow Area Using Boussinesq Models

Using a phase resolving model like a Boussinesq model the surface elevation is already in the time domain. In Fig. 2.5 snapshots of a shoaling waves passing circular towers are shown. These simulations are made by an arbitrary high-order finite-element method developed in [8]. The wave kinematics has, however, to be reconstructed from the used Boussinesq assumption. However, for the most frequently used enhanced Boussinesq models the wave kinematics is poor. The force on the circular tower can be computed by direct integration of the pressure on the submerged part of the structure. It may be better to couple the enhanced Boussinesq model to a local Reynolds-Average Navier Stokes (RANS) [9] model adjacent to the structure, taking the full viscous and free-surface characteristics of the problem into account.

2.7 Conclusions

It has been shown that for the purpose of assessing wave loads on wind-power plants in shallow or near shore sites in the Baltic, one can establish "deep-sea" wave climate by the Baltic WAM model, and transfer these waves by the phase averaging wave model SWAN to the more precise position of the plant. At this position non-linear wave kinematics can be realized in the time-domain by a second order realization and finally the wave loading be calculated by integrating Morison's equation to the instantaneous free surface. Alternatively the realized kinematics can be fed into a time-domain model e.g. an enhanced Boussinesq model or even a local RANS model.

2.8 Acknowledgements

The wave research is funded by the Swedish Energy Administration and FORMAS. The wind-energy application is done in co-operation with Risö

Forskningscenter and Teknikgruppen AB. The use of unpublished material from IBW Pan in Gdansk is acknowledged.

References

1. Komen, G. J. et al. (1994): *Dynamics and modelling of ocean waves*, Cambridge University Press, Cambridge, pp. 532
2. http://falowanie.icm.edu.pl/english/wavefrcst.html#elem2
3. Paplínska, B. (1999): Wave analysis at Lubiatowo and in the Pommeranian Bay based on measurements from 1997/1998 – Comparison with modelled data (WAM4 model), *Oceanologia*, 42(2), 241–254
4. Trumars, J. (2003): *Simulation of Irregular Waves and Wave Induced Loads on Wind Power Plants in Shallow Water*, Licentiate Thesis, Water Environment Transport, Chalmers University of Technology, Göteborg, 2003
5. Holthuijsen, L. H., Booij, N., Ris, R. C., Haagsma, I. G., Kieftenburg, A. T. M. M. and Kriezi, E. E. (2000): *User Manual, SWAN Cycle III version 40.11.* Delft: Delft University of Technology
6. Reda, A., and Paplínska, B. (2002): Application of the numerical wave model SWAN for analysis of waves in the surf zone, *Oceanological Studies*, XXXI(1–2), 5–21
7. Trumars, J., Tarp-Johansen, N. J. and Krogh, T. (2005): Fatigue loads on offshore wind turbines due to non-linear waves, *Proceedings of 24th OMAE2005*, June 12–17, Halkidiki
8. Eskilsson, C. and Sherwin, S. (2006): Spectral/*hp* discontinuous Galerkin methods for modelling 2D Boussinesq equations, *Journal of Computational Physics*, 212(2), 566–589
9. Ferziger, J. H. & Peric, M. (1997): *Computational Methods for Fluid Dynamics*, ISBN 3-540-59434-5, Springer, Berlin Heidelberg New York

3

Time Domain Comparison of Simulated and Measured Wind Turbine Loads Using Constrained Wind Fields

Wim Bierbooms and Dick Veldkamp

Summary. By means of so-called constrained simulation, wind fields can be generated which encompass measured wind speed series. If the method of constrained wind is used in load verification, the low frequency part of the wind and of the loads can be reproduced well, which makes it possible to compare time traces directly. However from the three load cases for the reference wind turbine investigated here, it appeared that there was no clear improvement in fatigue damage equivalent load ranges.

3.1 Introduction

Comparison between wind turbine load measurements and simulations is complicated by the uncertainty about the wind field experienced by the rotor. Usually the wind speed is measured at just one location (hub height) which makes direct comparison between the measured and simulation load time histories impossible. Instead random wind fields are generated (with the same mean and spectrum as the measured wind) and the power spectra of the loads or the equivalent fatigue load ranges are compared. The next section will present a method of so-called constrained stochastic simulation. By means of this method wind fields may be generated which encompass measured wind speed series i.e. one or more measured wind speed signals are reproduced exactly. Subsequently the constrained wind fields as well as the resulting wind turbine loading will be dealt with. Comparisons of the results of actual load measurements and simulated loads, found by using constrained wind fields based on measured wind, are given in Sect. 3.5.

3.2 Constrained Stochastic Simulation of Wind Fields

For wind turbine load calculations an artificial wind field is needed. Here we use the description by means of Fourier series:

$$u(t) = \sum_{k=1}^{K} a_{\mathrm{k}} \cos \omega_{\mathrm{k}} t + b_{\mathrm{k}} \sin \omega_{\mathrm{k}} t \tag{3.1}$$

The vector $u(t)$ represents the longitudinal velocity fluctuations in N points in the rotor plane (K frequencies). The Fourier coefficients a_{k} and b_{k} (row vectors) are independent for different frequencies and for the same frequency the following relation holds:

$$\mathrm{E}[a_{\mathrm{k}}\, a_{\mathrm{k}}^{\mathrm{T}}] = \mathrm{E}[b_{\mathrm{k}}\, b_{\mathrm{k}}^{\mathrm{T}}] = \frac{1}{T} S_{\mathrm{k}} \tag{3.2}$$

with T the total time of the sample and S_{k} the matrix of the cross-power spectral densities of the wind speed fluctuations in the different space points. They are found from the auto power spectral densities and the (root) coherence function γ.

In order to generate a spatial wind field we have to factorize the S_{k} matrix for each frequency component. In case the wind is measured at some location, it is straightforward to determine the set of Fourier coefficients for that space point. We can now improve the simulation of the wind speeds at the other space points on basis of the knowledge of the wind speed at that point. The accompanying equations can be found in the appendix of [1]. The wind fields which are simulated on basis of measured wind fields are called constrained stochastic simulations (they are generated under the constraint that at one or more points the measured wind speeds have to be reproduced). The generated wind fields may be considered to be wind fields which *could* have occurred during the measuring period; the actual wind field can never be reproduced in case of just a few measurement locations, since turbulence is a stochastic process.

3.3 Stochastic Wind Fields which Encompass Measured Wind Speed Series

With the method given in Sect. 3.2 several sets of wind time series are generated. As turbulence model the Kaimal spectrum and coherence function of IEC 61400-1 are chosen. Wind fields are created with mean wind speeds below, around and above rated wind speed of the reference turbines. Turbulence intensity is taken according to the IEC standard. For each situation at least 20 wind fields have been generated of 10 min length and a time step of 0.04 s. Nowadays it is common to generate all three velocity components; since the measurements (see Sect. 3.5) involved the longitudinal component only, the constrained simulations are restricted to that component.

The constrained simulations have been performed with two in-house packages. One uses a polar grid (30 azimuth by 5 radial, plus a point at the rotor

3 Time Domain Comparison Using Constrained Wind Fields

centre) and is used in combination with Flex5. The other applies a Cartesian grid (15 by 15) and produces wind fields which can be read by Bladed. In order to investigate the influence of the number of measuring points several configurations of anemometers (ranging from 1 to 11) have been considered. In Fig. 3.1 two of the considered configurations are shown.

One of the simulated wind fields can be appointed to be the "measured" wind field. The reduced scatter in the constrained wind time series is shown in Fig. 3.2. The shape of the time series gets more fixed (determined by the measured wind speeds). Unconstrained wind fields will average out to straight horizontal lines; the variance will be uniform over the rotor plane.

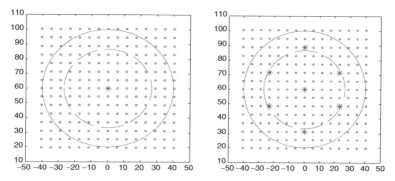

Fig. 3.1. The applied rectangular grid for the spatial wind fields and the locations of the anemometers of 2 of the considered configurations: 1 (left) and 7 (right); radius R and 2/3R are also indicated

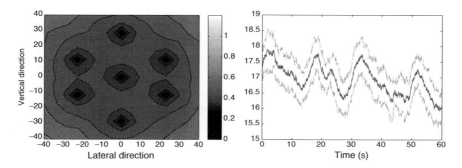

Fig. 3.2. Left: contour plot of the variance of the spatial constrained wind fields (averaged over time) over the rotor plane. Right: the mean (and plus/minus one standard deviation) of 30 constrained (7 anemometers) simulations; 17 m below the rotor centre

3.4 Load Calculations Based on Normal and Constrained Wind Field Simulations

The (un)constrained wind fields have been used as input for two standard wind turbine design tools: Flex5 and Bladed. Both packages feature among other things dynamic inflow, stall hysteresis and tip losses. Logarithmic wind shear is assumed and tower shadow is taken into account. In Flex5 the NEG Micon NM80/2750–60 is considered and in Bladed a generic 2 MW turbine. Both are pitch-regulated variable speed and have a diameter of 80 m, hub height of 60 m; the rated power is 2,750 kW and 2,000 kW respectively.

As example the time series of the calculated blade root flapping moment is shown in Fig. 3.3 based on the wind fields of Fig. 3.2. The response based on the "measured" wind is assigned "measured" load. The 1P (Ω=1.9 rad/s) response is clearly visible. The mean response of the constrained simulations gets more detailed, with increasing number of anemometers / constraints, and the standard deviation decreases since the wind fields gets more fixed. Comparison with the filtered "measured" load shows that the constrained response captures the low frequency phenomena well.

In order to consider fatigue, the damage equivalent load ranges are considered of the blade root flapping moment (S–N slope 12). The load ranges and the frequencies of occurrence are found with the usual rain flow counting procedure.

It is anticipated that constrained simulation results into a smaller scatter of the obtained equivalent damage. As measure for the scatter the COV (coefficient of variation) is taken. The convergence of the COV has been checked by doing 100 simulations for a particular situation. It turned out that about 20 simulations are enough; the uncertainty in the COV is about 0.1 to 0.2.

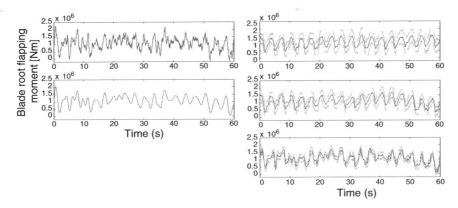

Fig. 3.3. Generic turbine: Left, top: "measured" flap moment; bottom: filtered flap moment. Right, from top to bottom: mean flap moment (plus/minus standard deviation) for 0, 1 and 7 anemometers/constraints. Note low frequency similarity

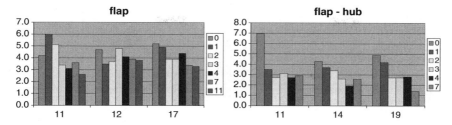

Fig. 3.4. COV of the blade equivalent flap moment range (left; generic turbine) and the flap moment range at the hub (right; NM80/2750). Mean wind speeds are indicated at bottom; legend indicates number of anemometers/constraints

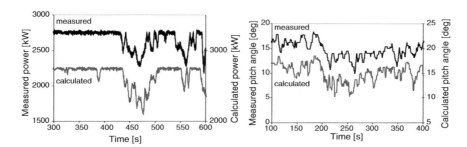

Fig. 3.5. Comparison between measurement and calculations (constrained wind in 1 point: rotor centre). Left: electrical power at $14.7\,\mathrm{ms}^{-1}$; right: pitch angle at $19.5\,\mathrm{ms}^{-1}$

In Fig. 3.4 (left) no clear trend can be observed; the reduction of the COV by constrained simulation is limited. Noticeable improvement is obtained however for the hub flap bending moment (right).

3.5 Comparison between Measured Loads and Calculated Ones Based on Constrained Wind Fields

Whether it is useful to use constrained wind in load verification can only be determined from a practical test. To this end we use the NEG Micon NM80 turbine in Tjæreborg (flat, open coastal terrain). The met-mast is located 200 m upwind of the turbine and has anemometers at 25, 41 and 60 m. Because the wind must come from a direction where the rotor sees the wind measured by the anemometer, only few measuring records could be used.

Two characteristic pictures are given in Fig. 3.5. It is clear that in both cases the low frequency behaviour is well captured (once the time delay from anemometer tower to wind turbine rotor has been incorporated, applying Taylor's frozen turbulence hypothesis), but that it is difficult to reproduce

W. Bierbooms and D. Veldkamp

Table 3.1. Relative mean equivalent flap moment range (calculated divided by measured; 100 = perfect correspondence) and relative standard deviation σ

Wind speed (ms^{-1})	11.4	14.7	19.5
0 constraints	87 (σ=4)	108 (σ=5)	107 (σ=4)
1 constraint	92 (σ=4)	103 (σ=4)	109 (σ=3)
3 constraints	92 (σ=3)	104 (σ=5)	112 (σ=3)

higher frequency phenomena. Even so, these comparisons are useful, for example it can be seen from Fig. 3.5 that apparently the pitch controller in the calculations is more "nervous" than the actual pitch system.

How good the load prediction is, is measured with the fatigue damage equivalent load. In Table 3.1 some results are given. Each selected measured 10 min. period was reproduced 20 times, respectively without constraint, and with wind constrained in 1 and 3 points. Mean and standard deviation of calculated load ranges are normalized by the measured load range. There is some improvement here and there, but on the whole it is not convincing. From the lack of improvement it must be concluded that the fatigue loads are mainly determined by high frequency wind variation, while the constraints only fix the low frequency wind (as can be seen in Figs. 3.3 and 3.5). We stress that at present it cannot be established whether this conclusion has general validity, because the limited number of load cases that were investigated.

3.6 Conclusion

If the method of constrained wind is used in load verification, the low frequency part of the wind and turbine loading can be reproduced well, which makes it possible to compare time traces directly. From the three load cases investigated here, it appeared that there was no improvement in predicted fatigue damage equivalent load ranges. However, more work is needed to verify this conclusion.

References

1. Wim Bierbooms, Simulation of stochastic wind fields which encompass measured wind speed series – enabling time domain comparison of simulated and measured wind turbine loads, EWEC, London, 2004

4

Mean Wind and Turbulence
in the Atmospheric Boundary Layer
Above the Surface Layer

S.E. Larsen, S.E. Gryning, N.O. Jensen, H.E. Jørgensen and J. Mann

Summary. Most wind energy related boundary layer formulations derive from the atmospheric surface boundary layer. The increasing height of modern wind turbines forces meteorologists to use relations more appropriate for greater heights. Such expressions are not as well established as the surface layer formulations. Data from the new Danish wind energy test site at Høvsøre is used to illustrate similarities and differences.

4.1 Atmospheric Boundary Layers at Larger Heights

For wind energy, the atmospheric boundary layer is normally considered surface layer with neutral thermal stability.

The surface layer wind profile can be written:

$$\overline{u(z)} = \frac{u_*}{\kappa} \left(\ln \left(\frac{z}{z_0} \right) - \psi \left(\frac{z}{L} \right) \right), \tag{4.1}$$

where u_* is the friction velocity, z the measuring height and z_0 the roughness length, κ the v. Karman constant, ψ the stability function and L is the length scale of thermal stratification:

$$\frac{z}{L} = -\frac{\overline{w\theta}}{u_*^3} \frac{g\kappa z}{T}, \tag{4.2}$$

with T and θ being temperature and potential temperature respectively. Neutral conditions correspond to $\psi(z/L \sim 0) \sim 0$, i.e. when z is small and or $|L|$ is large. Conversely for z large, L has to be even larger to ensure that the atmosphere can be considered neutral. For $|z/L| \gtrsim 0.5$ one will expect the stability term in (4.1) to be important. The scales, u_* and $\overline{w\theta}$, reflect the stress and heat-flux conditions at the surface.

The surface layer approximation of the atmospheric boundary layer demands that $z_0 \ll z \ll h$, where h denotes the boundary layer height. Since typically, $100 \lesssim h \lesssim 1,000\,\mathrm{m}$, the assumption becomes unrealistic with wind

22   S.E. Larsen et al.

turbine heights larger than, say $100\,\mathrm{m}$. Therefore one has to account for that the eddy size is limited by h, which must enter into the formulations, supplementing the length scales, z_0, z, L already being present [1].

One way of having h entering the formulations is to use local scaling, as opposed to the formulation in (4.1 and 4.2) above. Here, often-used approximations are [2]:

$$
\begin{aligned}
\frac{u_{*,l}}{u_{*,0}} &= \left(1 - \frac{z}{h}\right)^{\alpha/2} \text{ with } \alpha = 1 - 2 \\
\frac{\overline{w\theta_l}}{\overline{w\theta_0}} &= \left(1 - \frac{z}{h}\right)^{\beta} \text{ with } \beta \approx 1 - 3,
\end{aligned}
\tag{4.3}
$$

where we have taken subscript 0 to refer to surface values and l to local values, meaning at height, z. Local scaling has often been found to work well with data when larger heights or heterogeneous conditions are involved [3].

For neutral–stable conditions, also, the stability length, L, must be supplemented with other scales, reflecting the thermal characteristics of the atmosphere. The most important for stable- or neutral conditions is derived from the Brunt–Vaisala frequency, N, of the free atmosphere:

$$
L_{\mathrm{N}} = \frac{u_*}{\mathrm{N}}, \ \mathrm{N}^2 = \left(\frac{\mathrm{g}}{T}\frac{\partial \theta}{\partial z}\right)_{z>h}.
\tag{4.4}
$$

L_{N} reflects the role of the stability of free atmosphere for the growths of the boundary layer, and thereby for its largest eddies. Several methods exist for combination of the various length scales. For the L-scales, the following method can be applied [4]:

$$
L_{\mathrm{C}}^{-1} \approx \left(L^{-2} + L_{\mathrm{N}}^{-2}\right)^{\frac{1}{2}}.
\tag{4.5}
$$

The length scale, L_{C}, replaces L in boundary layer equations, including the ones presented in this paper, and is most important for neutral: $L \sim \infty$.

For changing surface conditions the boundary layer develops internal boundary layers reflecting the new surfaces. These IBLs generally develops slower and slower as the fetch and thereby the height increases. Therefore greater heights will tend to reflect surface conditions upstream and the farther upstream the larger the measuring height [5].

4.2 Data from Høvsøre Test Site

The Høvsøre test site for large wind turbines is situated at the Danish North Sea coast, with one $100\,\mathrm{m}$. climatology mast, and two $165\,\mathrm{m}$ warning light masts. The site is situated about $1{,}500\,\mathrm{m}$. from the shoreline, meaning that for flows from West, measurements for $z >$ about $50\,\mathrm{m}$ can be considering

4 Mean Wind and Turbulence in the Atmospheric Boundary Layer

as representing the marine atmosphere. Easterly winds on the other hand represent a terrestrial boundary layer.

In Fig. 4.1 we have plotted the annual stability distributions for easterly and westerly flows. The figures both reflect the tendency to stable stratification at larger heights and that thermal effects have to be included also for the winds of relevance for wind energy.

Next we plot in Fig. 4.2, the annual mean wind profiles from East and West for $u_{100} > 8, 10, 12, 14, 16 \, \text{ms}^{-1}$, and for z between 10 and 160 m. From West the internal boundary layer growth is clear in the kinks in the profiles and in agreement with the standard models [5]. Most from East but also from West the tendency for the upper part of the profiles to increase more than a logarithmic law is clear and can be modeled to be in accordance with (4.1) to that of (4.5) for climatologic value of $N \sim 0.01 \, \text{s}^{-1}$, yielding $L_N \sim 2\text{--}300 \, \text{m}$.

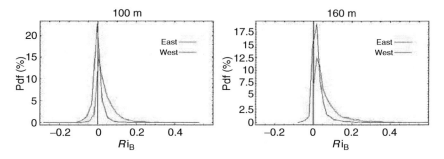

Fig. 4.1. Annual stability distribution for $u_{100} > 8 \, \text{ms}^{-1}$ from East (terrestrial) and West (marine) boundary layers at $z=100$ and 160 m above terrain. The width of the PDF corresponds to a Bulk Richardson number, Ri_B with $\Delta\theta$ defined between 80–100 and 100–165 m respectively, of about $\pm\ 0.02$ for both figures. This corresponds to $|100/L| \sim 1.3$ over land and 3.4 over water. Note, $Ri_B = (g/T)(\Delta\theta/\Delta z)(z/u_z)^2$

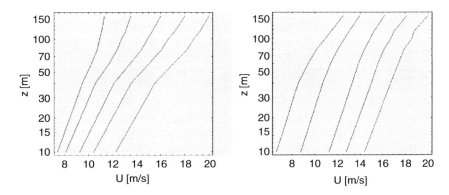

Fig. 4.2. Profiles of wind from West (left) and East (right hand side), for 10–160 m over terrain, see text above for discussion.

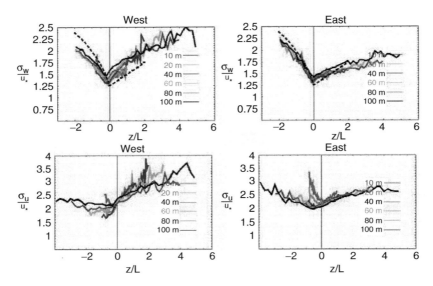

Fig. 4.3. Scaled standard deviations: σ_w/u_* and σ_u/u_* vs. z/L for $z \sim 10$–100 m for West and East flows. Local scaling has been used, see (4.3). Broken lines: Analytical forms [2]

In Fig. 4.3 we present finally the scaled standard deviations of velocity, σ_w/u_* and σ_u/u_* vs. z/L, for z between 10 and 100 m. For σ_w/u_* analytical expressions are shown as a broken line. No general simple form exists for σ_u/u_* vs. z/L. For unstable conditions it becomes a function of h/L and for stable condition it growths slowly with z/L. We believe that use of local scaling is partly responsible for the some of behavior with z/L since h varies somewhat systematically with z/L.

4.3 Discussion

We have considered mean flow and turbulence over land and water based on data from Høvsøre, above the surface layer proper. We have seen that thermal stability cannot be neglected for these heights, neither over land nor water, that the wind profiles adapt to the standard models for internal boundary layers, and that the wind aloft has a tendency to increase more than logarithmic, either due to the boundary length scale of [1] or due to the influence of stability and a length scale based on the Brunt–Vaisala frequency [4]. For neutral to stable conditions it is understood that these two scales may be related.

Additionally, we have found that the scaled σ_w/u_* vs. z/L largely follows standard formulations, while σ_u/u_* is less well behaved. Some of the behavior of σ_w/u_* and σ_u/u_* can be explained by that the plots are based on local scaling, and that the boundary layer height changes somewhat systematically with z/L in the plots. For unstable conditions, σ_u/u_* is known to depend on h/L rather than on z/L, where h is the boundary layer height. This is likely to explain the scatter for σ_u/u_* for unstable condition.

We have not used h directly as a scaling parameter in this paper because it is presently not measured. For the wind speed profiles we have not discussed the unstable profiles much, but they are well described by the standard profile formulations, surface layer leading to free convection and mixed layer formulations as the height increases.

References

1. H. Panofsky. Tower Micrometeorology. In Workshop on Micrometeorology. Am. Met. Soc., 151–176, 1972
2. R.B. Stull. An introduction to Boundary Layer Meteorology, Kluwer, 1988
3. D. Vickers and L. Mahrt. Observations of non-dimensional wind shear in a coastal zone. Q.J.R. Meteorol. Soc., 125, 2685–2702, 1999
4. S. Zilitinkevich and I.N. Esau. Resistance and heat-transfer laws for stable and neutral planetary boundary layers: Old Theory advanced and revaluated. Q.J.R. Meteorol. Soc., 206, 131, 1863–1893, 2005
5. A.M. Sempreviva, S.E. Larsen, N.G. Mortensen, and I. Troen. Response of neutral boundary layers to changes of roughness. BoundaryLayer Meteorol., 50, 205–225, 1990

5

Wind Speed Profiles above the North Sea

J. Tambke, J.A.T. Bye, B. Lange and J.-O. Wolff

To achieve reliable wind resource assessments, to calculate loads and wakes as well as for precise short-term wind power forecasts, the vertical wind profile above the sea has to be modelled for tip heights up to 160 m. In previous works, we analysed marine wind speed profiles that were measured at the two met masts Horns Rev (62 m high) and FINO1 (103 m) in the North Sea (e.g. [3]). It was shown that the wind shear above 50 m height is significantly higher than expected with Monin–Obukhov corrected logarithmic profiles, revealing an almost linear increase of the wind speed. For that reason, we developed a new analytic model of marine wind velocity profiles, which is based on inertial coupling between the Ekman layers of the atmosphere and the ocean. The good agreement between our theoretical profiles and observations at Horns Rev and FINO1 support the basic assumption in our model that the atmospheric Ekman layer begins at 15 to 30 m height above the sea surface.

5.1 Theory of Inertially Coupled Wind Profiles (ICWP)

The geostrophic wind is regarded as the driving force of the wind field in lower layers of the atmosphere. The momentum is transferred downwards through the Ekman layers of atmosphere and ocean, which are modelled with a constant turbulent viscosity. The challenge is to derive an adequate description of the coupling between the two Ekman layers, where the idea is to introduce a connecting wave boundary layer with a logarithmic wind profile that reaches only up to a maximal height of 30 m.

In order to derive the coupling relations between the three layers the following similarity assumptions are made. First, close to the surface the ratio between the drift velocities of air, u_{air}, and water, u_{water}, as well as the ratio between the friction velocities is given by $\frac{u_{\mathrm{water}}}{u_{\mathrm{air}}} = \frac{w_*}{u_*} = \sqrt{\frac{\rho_{\mathrm{air}}}{\rho_{\mathrm{water}}}} \approx \frac{1}{29}$,

28   J. Tambke et al.

where u_* is the friction velocity of the air flow and w_* the one of the water flow while ρ_{air} and ρ_{water} are the respective densities. This is equivalent to assuming that the shear stress is constant across the interface between air and water. The constant stress defines the wave boundary layer, extending from a height z_{B} above to $-z_{\text{B}}$ below the water level. This wave boundary layer is similar to the surface layer onshore where the logarithmic wind profile is valid.

Second, the inertial coupling relation [1] represents the shear stress in the wave boundary layer in the form of a drag law with regard to the inertially weighted fluid speeds at the limits of this layer, where K_{I} is a specific drag coefficient: $\tau_{\text{wave}} = K_{\text{I}}[\sqrt{\rho_{\text{air}}}\, u(z_{\text{B}}) - \sqrt{\rho_{\text{water}}}\, u(-z_{\text{B}})]^2$.

Third, analogous to the first similarity relation the turbulent viscosities ν_{air} and ν_{water} of the two Ekman layers are also assumed to be weighted according to the inverse density ratio of the two fluids: $\frac{\nu_{\text{water}}}{\nu_{\text{air}}} = \frac{\rho_{\text{air}}}{\rho_{\text{water}}}$.

These conditions allow the determination of the velocities of the two fluids at certain heights as an implicit function of a given geostrophic wind. The full profiles can be derived in the following way: Due to the constant shear stress the wind profile in the wave boundary layer has a logarithmic shape for neutral thermal stratification. Non-neutral conditions can be modelled with an additional stability function Ψ_{m} and the Monin–Obukhov (MO) length L. Expressed in a coordinate system where the horizontal component of the stress tensor $\boldsymbol{\tau}_h$ is parallel to the x-axis the profile can be written as

$$\boldsymbol{u}(z) = \left(u_{\text{L}} + \frac{u_*}{\kappa} \ln\left(\frac{z}{z_{\text{R}}} \right) + \frac{u_*}{\kappa} \Psi_{\text{m}}\left(\frac{z}{L} \right), v_{\text{L}} \right), \tag{5.1}$$

and is valid for $z_{\text{R}} \leq z \leq z_{\text{B}}$; κ is the von Karman constant and z_{R} is the theoretical height where the momentum transfer from the air to the wave field is centred. The above mentioned drift velocities are $u_{\text{air}} = u(z_{\text{R}}) = u_{\text{L}}$ and $u_{\text{water}} = u(-z_{\text{R}}) \approx \frac{1}{29} u_{\text{L}}$. As a result of the above assumptions and the choice of the coordinate system the drift $(u_{\text{L}}, v_{\text{L}})$ at the very small height z_{R} is collinear to the geostrophic wind: $(u_{\text{L}}, v_{\text{L}}) = \frac{1}{2}\mathbf{G}$, where $\mathbf{G} = (u_{\text{g}}, v_{\text{g}})$. This theoretical, though curious relation has no relevance in-situ, since neither $(u_{\text{L}}, v_{\text{L}})$ close to the waves nor $\mathbf{G}$ can be measured directly. Note that in the chosen coordinate system $u_{\text{g}} > 0$ and $v_{\text{g}} < 0$.

Concluding from [2], the relation that connects the friction velocity u_* at the water surface to the geostrophic wind is given by

$$|\mathbf{G}| = \frac{\sqrt{r^2 + 1}}{|r + 1|} \frac{u_*}{\sqrt{K_{\text{I}}}}. \tag{5.2}$$

The angle of rotation of the surface stress tensor to the left hand side of the geostrophic velocity (in the northern hemisphere) is atan(-1/r) with $-1 < r < -\infty$, where r still has to be determined.

The height of the wave boundary layer z_B can be related to the wave field. Previous results from oceanography suggest that z_B is reciprocal to the peak wave number, k_p, of the wave spectrum [2]. The corresponding peak wave velocity, c_p is given by $c_p = \sqrt{g/k_p}$ where g is the gravitational constant. Assuming that c_p is proportional to the wind speed $u(z_B)$, i.e. $c_p = B\,u(z_B)$, the height of the wave boundary layer can be written as

$$z_B = \frac{B^2}{8\,g} \frac{(2r+1)^2}{r^2+1} \, |\mathbf{G}|^2. \tag{5.3}$$

For heights above z_B the velocity vector in the Ekman layer is described by the well-known Ekman spiral

$$u(\hat{z}) = \hat{u}_1 \left(\cos(\beta\hat{z}) - \sin(\beta\hat{z})\right) \mathrm{e}^{-\beta\hat{z}} + u_g$$
$$v(\hat{z}) = \hat{v}_1 \left(\cos(\beta\hat{z}) + \sin(\beta\hat{z})\right) \mathrm{e}^{-\beta\hat{z}} + v_g \tag{5.4}$$

with $\hat{z} = z - z_B$ and $\beta = \sqrt{f/(2\hat{\nu})}$ where f is the Coriolis parameter, $\hat{\nu}$ the turbulent viscosity and $\hat{u}_1 = 0.5\,u_g/r$ and $\hat{v}_1 = -0.5\,v_g$.

At the interfaces between wave boundary layer and Ekman layer, $z = \pm z_B$, the turbulent stress tensor and the turbulent viscosity is assumed to be continuous such that the wind profiles can be matched smoothly. This matching condition at the transition between wave boundary layer and Ekman layer provides the necessary link to calculate the parameter r. At $z = z_B$ the height dependent viscosity $\nu(z_B) = \kappa\,u_*\,z_B\,\Phi_\mathrm{m}$ of the wave boundary layer, where Φ_m is a MO-parameter, and the constant viscosity $\hat{\nu} = \frac{2}{f}(r+1)^2\,K_I\,u_*^2$ of the Ekman layer are set equal, $\nu(z_B) = \hat{\nu}$. Using (5.2) and (5.3) this provides the defining equation for r:

$$-\kappa\,\Phi_\mathrm{m}\,|\mathbf{G}|\,\sqrt{K_I} \left(\frac{B}{4K_I}\right)^2 \left(\frac{f}{g}\right) \frac{(2r+1)^2}{(r+1)^3\sqrt{r^2+1}} = 1. \tag{5.5}$$

The parameters $K_I = 1.5 \cdot 10^{-3}$ and $B = 1.3$ were obtained from independent oceanographic measurements [2], f and g are known constants at the desired location. Φ_m has to be given or set to 1 for neutral conditions. The geostrophic wind $\mathbf{G}$ as driving force can be chosen to match a given wind speed at a given height. Hence, (5.5) can be iteratively solved for r and the vertical wind profiles according to (5.1) and (5.4) can be calculated.

5.2 Comparison to Observations at Horns Rev and FINO1

The wind speed profiles according to the ICWP theory and to the standard offshore WAsP profile and to other standards are calculated with the time

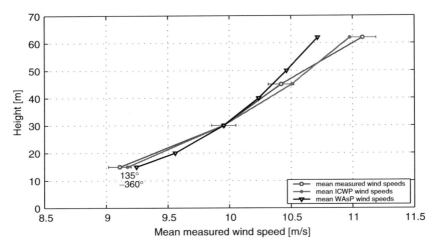

Fig. 5.1. Mean profiles at Horns Rev for the undisturbed sector 135–360°, $u > 4\,\mathrm{m\,s^{-1}}$

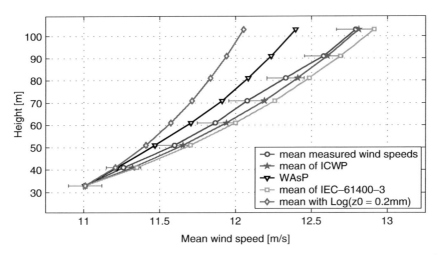

Fig. 5.2. Mean profiles at FINO1 for the undisturbed sector 190–250°, $u > 4\,\mathrm{m\,s^{-1}}$

series of wind speeds measured at 30 m height as input. Figures 5.1 and 5.2 show the resulting mean profiles with very small biases of the ICWPs. Figure 5.3 reveals that the non-logarithmic shape of the observed profiles occurs in all conditions of thermal stratification, as it was shown for Horns Rev in [3]. The large range of wind speed ratios due to varying thermal stratification lead to a root mean square error (RMSE) of the modelled wind speeds of 5% (FINO1) and 6% (Horns Rev) at 62 m height, and of 10% at 103 m. When

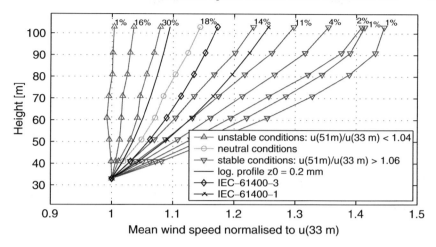

Fig. 5.3. Profiles and frequencies of different classes of thermal conditions at FINO1

we use the single wind shears between 33 and 51 m height to calculate Φ_m in order to provide the ICWP-model with information about the stability of the atmospheric stratification, the RMSE is reduced to 3.6% at 103 m height.

References

1. J.A.T. Bye. Inertial coupling of fluids with large density contrast. *Phys. Lett. A*, **202**:222–224, 1995
2. J.A.T. Bye. Inertially coupled Ekman layers. *Dyn. Atmos. Ocean*, **35**:27–39, 2002
3. J. Tambke, J.A.T. Bye, M. Lange, U. Focken, and J.-O. Wolff. Forecasting Offshore Wind Speeds above the North Sea. *Wind Energy*, **8**:3–16, 2005

6

Fundamental Aspects of Fluid Flow over Complex Terrain for Wind Energy Applications

José Fernández Puga, Manfred Fallen and Fritz Ebert

Summary. The flow over steep hills has been studied experimentally and numerically. The sinusoidally-shaped hills follow the parametric shape equations of the RUSHIL experiments. In the wind tunnel of the University of Kaiserslautern the flow over the hill was measured with the Laser Doppler Anemometry. Thereby, devices had to be found in order to thicken artificially the boundary layer. The numerical investigation has been carried out using several turbulence models comparing the calculated results with the experimental ones. In order to carry out reliable wind energy power predictions recommendations concerning numerical calculations are given.

6.1 Introduction

Wind speed and power predictions are of high importance when installing new wind parks, even more in complex terrain. The topographic effects on air flow cause speed-ups and increase velocity fluctuations which have a significant influence on wind energy resource assessment and on wind loading on structures. As sites in forests are becoming interesting the flow field in and above the canopy should also be taken into account [1].

Wind speed and power are mostly forecasted using linearized models which do not count very well for the topographic effects. Moreover, measurements at low heights are used in order to interpolate data. The forecast is therefore of low accuracy. Though being expensive, the best method for predicting wind power is measuring the velocities on site at hub height. Additionally, the measured velocities must be correlated with a long-term measurement of at least 20 years in order to define whether the measured period is an average wind year or not [2].

The present investigation has been focused on the flow over complex terrain, especially on flow over steep hills. Though being important stratification has been neglected focusing on a neutral flow.

6.2 Experimental Setup

The measurements have been carried out in the wind tunnel of the University of Kaiserslautern using the laser Doppler anemometry operating in backscatter mode. The seeding particles which have a size of 2μm are made by spraying a solution of glycerine and water.

In order to carry out feasible measurements the experimental setup has been chosen according to the German VDI-guideline 3783. The boundary layer has been thickened artificially using spires which were fixed at the entrance of the test section. The power law exponent of the modelled atmospheric boundary layer (Fig. 6.1) has a value of 0.16 which corresponds to a moderately rough terrain according to the guideline.

Furthermore, the turbulence intensities in longitudinal, lateral and vertical direction (Fig. 6.2) represent also a moderately rough terrain. The spectral

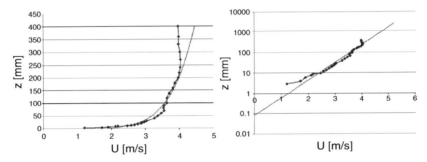

Fig. 6.1. Modelled atmospheric boundary layer

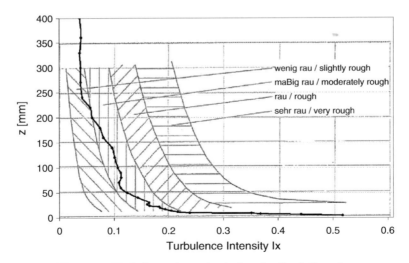

Fig. 6.2. Turbulence intensity in longitudinal direction

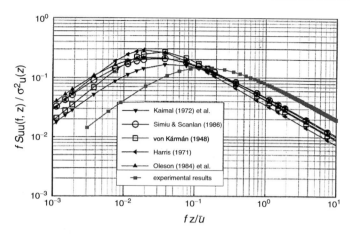

Fig. 6.3. Spectral density distributions of the kinetic energy of turbulence

Table 6.1. Dimensionless recirculation lengths

	Recirculation length ($\frac{L}{W}$)
Measurement	0.63
$k - \epsilon$ RNG	0.66
$k - \omega$ SST	0.60
LES-SL	0.64

density distribution shows a maximum at higher frequencies in comparison to the distributions given in literature (Fig. 6.3).

The three-dimensional hill model has been mounted on a flat plate which extends over the entire length and width of the test section. The hill has a height H of 100 mm and a half width $\frac{W}{2}$ of 250 mm which leads to a characteristic half-width to height ratio of 2.5. The measurements have been carried out with a free stream velocity of $4\frac{m}{s}$.

6.3 Results

The numerical calculations have been performed with a commercial unstructured finite-volume solver which employs a collocated grid. For all equations a second-order upwind discretization scheme was used, while the pressure-velocity coupling has been done with the SIMPLE method. Several turbulence models and some of their variations have been employed [3]. The grid has been refined up to the wall reaching dimensionless wall distances of $y^+ \simeq 4 - 11$ in order to be able to represent the recirculation zone in the lee of the hill.

The $k - \epsilon$ RNG and the $k - \omega$ SST turbulence model calculate the length of the recirculation zone in the lee of the hill in good agreement with the measurement (Table 6.1). Both models perform well calculating mean velocity values but differ remarkably regarding turbulence values. Best agreement is achieved performing a large-eddy simulation. The profiles for the mean velocity and the turbulent kinetic energy fit very well to the measured profiles

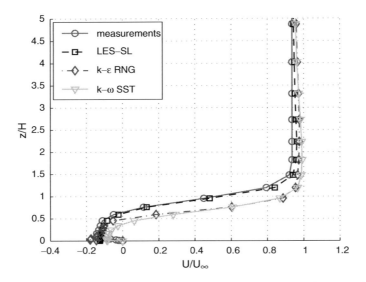

Fig. 6.4. Longitudinal mean velocity in the lee of the hill ($\frac{x}{W} = 0.5$)

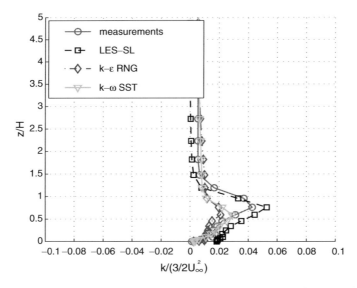

Fig. 6.5. Turbulent kinetic energy in the lee of the hill ($\frac{x}{W} = 0.5$)

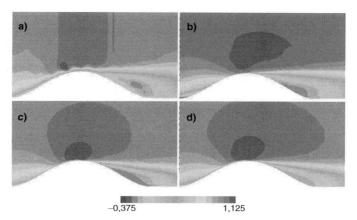

Fig. 6.6. Contour plot of the longitudinal mean velocity U/U_∞; (a) measurement, (b) LES-SL, (c) k–ϵ RNG, (d) k–ω SST

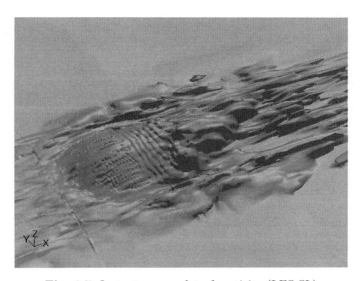

Fig. 6.7. Instantaneous plot of vorticity (LES-SL)

(Figs. 6.4 and 6.5). Moreover, the size of the recirculation zone in the lee of the hill is calculated best (Fig. 6.6). Additionally, the unsteadiness of the flow becomes evident regarding vorticity (Fig. 6.7).

6.4 Conclusions

The knowledge of the atmospheric flow over hills is of high importance in wind energy purposes. In order to predict the energy production rate of wind energy converters numerical calculations of the flow field are necessary when no measurements are available. Due to the fact that the flow phenomena in the atmosphere are turbulent the choice of the right turbulence model is essential. It has been shown that the velocity field has been calculated in agreement with the measurements in the wind tunnel of the University of Kaiserslautern using the $k - \epsilon$ RNG and the $k - \omega$ SST turbulence model as well as a second-order discretization scheme and a refined near-wall mesh. Best agreement is achieved performing a large-eddy simulation.

References

1. Fernández Puga J, Fallen M, Ebert F (2003) Wind energy converter siting in forest landscapes. In: Conference Proceedings of EWEC 2003. Madrid
2. Fernández Puga J, Fallen M, Ebert F, Tetzlaff G (2002) Long-term prediction of wind resources by use of a mean wind year. In: Conference Proceedings of the World Wind Energy Conference and Exhibition. Berlin
3. Fernández Puga J, Fallen M, Ebert F (2004) Evaluation of turbulence models for wind energy production predictions. In: Conference Proceedings of EWEC 2004. London

7

Models for Computer Simulation of Wind Flow over Sparsely Forested Regions

J.C. Lopes da Costa, F.A. Castro and J.M. L.M. Palma

7.1 Introduction

Because of the flow complexity and wind power reduction, the regions near or within forests have not been considered for wind park installation. However, because of the shortage of good clear sites, this situation was reverted, adding new motivation to the study of the wind flow over forested regions.

The so-called $k - \varepsilon$ model, the standard in flow modelling of engineering applications, has also appeared in atmospheric flow studies, circumventing the limitations of simpler approaches [1]. However, in the case of canopy flows, there are open questions on the actual performance of the $k - \varepsilon$ compared with lower order models, and the mathematical form of the additional terms in the transport equations for the turbulence kinetic energy k and its rate of dissipation ε [2].

7.2 Mathematical Models

The canopy models as in [4, 6] and a version of $k - \varepsilon - \overline{v^2} - f$ [3] are studied over model forests. For sake of compactness, the model description will be restricted to their differences from the standard versions, i.e. without the terms accounting for the canopy. The reader may refer to the original publications for a complete description of each of the models being used here.

The presence of the canopy requires an additional term in the momentum equation accounting for the extra drag, and re-definition of the production terms of k and ε in the transport equations, where additional terms S_k and S_ε will appear $(\mathcal{P}_k - \rho\varepsilon + S_k)$ and $(C_1\varepsilon/k\mathcal{P}_k - C_2\rho\varepsilon^2/k + S_\varepsilon)$. In case of $k - \varepsilon - \overline{v^2} - f$ turbulence model, the equation for f contains also an additional S_k, $C_{f2}(\mathcal{P}_k + S_k)/k$.

The terms S_k and S_ε will read

$$S_k = \frac{1}{2}\rho\alpha C_D \left(\beta_p \left| \overline{\mathbf{U}} \right|^3 - \beta_d \left| \overline{\mathbf{U}} \right| k \right), \tag{7.1}$$

$$S_\varepsilon = \frac{1}{2}\rho\alpha C_D \left(C_{4\varepsilon}\beta_p \frac{\varepsilon}{k} \left|\mathbf{U}\right|^3 - C_{5\varepsilon}\beta_d \left|\mathbf{U}\right|\varepsilon \right), \tag{7.2}$$

where C_D and α are the tree drag coefficient and density foliage, and the closure constants are defined according with the model being used.

Model	β_p	β_d	$C_{4\varepsilon}$	$C_{5\varepsilon}$
Svensson et al. [6]	1.0	0.0	1.95	–
Liu et al. [4]	1.0	5.1	1.5	0.4
$k - \varepsilon - \overline{v^2} - f$ [3]	1.0	6.75	1.5	1.5

Note that when $C_{4\varepsilon} = C_{5\varepsilon}$, the S_ε formulation becomes

$$S_\varepsilon = C_{4\varepsilon}\frac{\varepsilon}{k}S_k, \tag{7.3}$$

which corresponds to the formulation in [6]. For simplicity, the equations above were written in Cartesian coordinates, whereas the code is in general coordinates, appropriate for computer simulation of complex terrain, cf. [1].

The ground surface was modelled by wall laws. The mean and turbulent fields at the inflow boundary followed from horizontally homogeneous atmospheric boundary layers. The velocity was tangential at the top and lateral boundaries. At the outflow boundary the velocity was obtained by linear extrapolation from the inner nodes, constrained by global mass conservation. The pressure at the boundaries was obtained by linear extrapolation from the inner nodes. At the lateral and top boundaries the turbulent quantities were obtained by linear extrapolation.

7.3 Results

Computer simulations of the flow over a forested ($C_D = 0.8$, $\alpha = 6.25\,\mathrm{m}^{-1}$) sinusoidal hill and along a flat clearing, downstream of a forest ($C_D = 0.3$ and α between 6.0 and $57.5\,\mathrm{m}^{-1}$) region are compared with wind tunnel measurements [4, 5].

The results are a sample of an ongoing appraisal of turbulence models of flow over forested regions, which we are showing here with the main purpose of illustrating the difficulties that one is currently faced with.

Flow over a modelled forested sinusoidal hill

The trees can account for an increase of both the turbulence and its dissipation rate, yielding lower turbulence within the canopy height and higher turbulence values above the tree top. This is the trend in Fig. 7.1, for both the experimental data and the model results.

Downstream of the hill top, the k distribution along the vertical, and the location of the maximum, is set by the mean vertical shear, due to the mixing

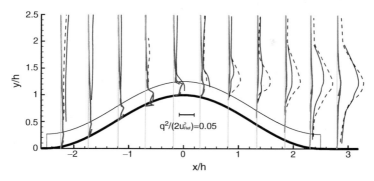

Fig. 7.1. Turbulent kinetic energy over a forested hill. Wind tunnel data (*dashed line, solid line*); Svensson [6] (*dashed line*), Liu [4] (*solid line*)

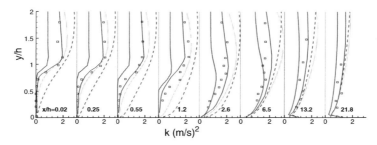

Fig. 7.2. Turbulence kinetic energy along a forest clearing. Wind tunnel data (□); Svensson [6] (*dashed line*), Liu [4] (*solid line*), v2-f model (*dotted line*)

layer type of flow above the trees. The results by the two models in [4, 6] (7.2 and 7.3) are virtually identical, and downstream of the hill, 2–5 tree heights above the top of the trees, the difference between wind tunnel data and model results is too high. The experimental and computer result differ by a factor of 10, suggesting hidden aspects in the experimental data that are beyond model capabilities. For instance the presence of large unsteady flow structures originating from the trees at hill top, which would have gone unnoticed by a conventional averaging of the LDA measurements.

Flow over a modelled flat forest

The effect of the two alternative formulations as in [6] and [4], is most obvious in the case of the wind flow along a clearing, Fig. 7.2. Model [6] yields turbulence values which are too high everywhere, as a result of insufficient turbulence dissipation along the forest upstream of the clearing. The two laboratory models (sinusoidal hill and flat clearing) differ also in terms of the forest density, which may evidence the differences between the model approaches.

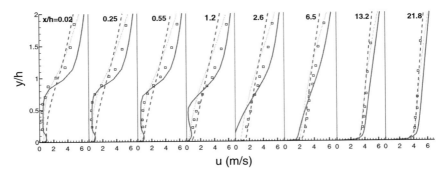

Fig. 7.3. Longitudinal velocity along a forest clearing. Wind tunnel data (*dashed line, solid line*); Svensson [6] (*dashed line*), Liu [4] (*solid line*), v2-f model (*dotted line*)

Concerning the mean velocity (Fig. 7.3), all models show a much better agreement with the wind tunnel data, compared with Fig. 7.2.

7.4 Conclusions

There are difficulties in the computer simulation of wind flow over forests, related with model parameterisation, which require the availability of good experimental data.

References

1. F. A. Castro, J. M. L. M. Palma, and A. Silva Lopes. Simulation of the Askervein flow. Part 1: Reynolds averaged Navier-Stokes equations (k-ϵ turbulence model). *Boundary-Layer Meteorol.*, 107:501–530, 2003
2. G. G. Katul, L. Mahrt, D. Poggi, and Christophe Sanz. One- and two-equation models for canopy turbulence. *Boundary-Layer Meteorol.*, 113:81–109, 2004
3. F. S. Lien and P. A. Durbin. Non-linear k-ε - v modeling with application to high-lift. In *Proc of the Summer Program*. CTR, Stanford University, 1996
4. J. Liu, J. M. Chen, T. A. Black, and M. D. Novak. E-ε modelling of turbulent air flow downwind of a model forest edge. *Boundary-Layer Meteorol.*, 77:21–44, 1996
5. B. Ruck and E. Adams. Fluid mechanical aspects in the pollutants transport to coniferous trees. *Boundary-Layer Meteorol.*, 56:163–195, 1991
6. U. Svensson and K. Häggkvist. A two-equation turbulence model for canopy flows. *J. Wind Eng. Ind. Aerodyn.*, 35:201–211, 1990

8

Power Performance via Nacelle Anemometry on Complex Terrain

Etienne Bibor and Christian Masson

8.1 Introduction and Objectives

In order to improve the performance of wind turbines (WT) and to reduce discrepancies between concerned parties, testing standards have been developed. Those proposed by the International Electrotechnical Commission (IEC 61400-12) constitute an international reference. However, several aspects treated in these standards are not perfectly well understood or are based on inappropriate assumptions. For example, the power performance verification of a wind park on complex terrain raises significant difficulties related to the technique of nacelle anemometry [1]. The goal of the present study is to evaluate the precision of this technique on complex terrain.

8.2 Experimental Installations

To perform this study, experimental installations have been deployed on a very complex site in Eastern Canada. Three WTs are present at the Riviere-Au-Renard Wind Farm (PER). The WT have a nominal power of 750 kW, with a rotor diameter of 48 m and a hub height of 46 m. They are active control WT, with variable speed and fixed pitch angle. The rotational speed, with a range between 8.5 and 24.8 rpm, is adjusted in order to operate as often as possible at the optimal point, $C_{\mathrm{P,max}}$. This $C_{\mathrm{P,max}}$ of 0.425 occurs approximately at a tip speed λ of 7.2 ($\lambda = \Omega R/V_\infty$, where Ω represents the rotational speed and R is the rotor radius)

8.3 Experimental Analysis

An experimental analysis has been undertaken. The first step was to perform a site calibration. For every valid sectors of $10°$, a correlation between $V_{\mathrm{ref}} \Leftrightarrow V_\infty$

44   E. Bibor and C. Masson

was established. Treatment of data required important cares during the different steps: determination of valid sectors and filtration of the data. Nacelle anemometry's correlation $V_{\mathrm{nac}} \Leftrightarrow V_\infty$ was then calculated by synchronizing the data from the WT and the reference tower.

8.4 Numerical Analysis

A numerical analysis as been performed with the objective of having a better understanding of nacelle anemometry. In order to be as realistic as possible, a turbulent flow was simulated. For that purpose, the Reynolds-Average Navier-Stokes (RANS) equations were solved. The $k - \epsilon$ turbulence model was used to close the system of equations. The rotor is represented as an actuator-disc surface [2], on which external forces are prescribed according to the blade-element-theory.

The commercial software FLUENT was chosen to perform this numerical analysis. The rotor can be modelled by a pressure jump calculated from the forces found previously. Also, a two-dimensional axisymmetric flow was studied in order to benefit from the symmetrical geometry of the nacelle. The size of the calculation domain was optimized so that the results were not influenced by the boundary conditions, but also that the computing time was optimum.

8.5 Results and Analysis

8.5.1 Comparaison with the Manufacturer

As explained in Sect. 8.3, a nacelle anemometry correlation $V_{\mathrm{nac}} \Leftrightarrow V_\infty$ was experimentally obtained: $V_\infty = 0.953 + 0.791\, V_{\mathrm{nac}}$. The latter was compared with the correlation provided by the manufacturer: $V_\infty = 0.536 + 0.723\, V_{\mathrm{nac}}$.

For $V_{\mathrm{nac}} = 20\,\mathrm{m/s}$, differences up to $1.59\,\mathrm{m/s}$ were measured. In order to determine the validity of those results, power curves were constructed using the two correlations. The use of the manufacturer's correlation leads to unrealistic results, with power coefficients (C_P) higher than the theoretical limit of 0.475 for this WT. These results confirm the inappropriateness in using a common correlation for all WTs of the same type, with no consideration of the type of terrain.

8.5.2 Influence on the Wind Turbine Control

The errors made in the estimation of V_∞ have important consequences on the performances of the WT. Indeed, the active control WT continuously modifies his rotational speed in order to operate at the optimum tip speed ratio $\lambda_{\mathrm{opti}} = 7.2$. An error in V_∞ will automatically lead to an error in λ calculated by the controller. Thus, the WT will always operate at C_Ps lower than $C_{\mathrm{P,max}}$.

8 Power Performance via Nacelle Anemometry on Complex Terrain   45

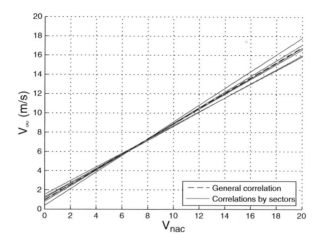

Fig. 8.1. Comparison between correlations by sectors and general correlation for all valid sectors on correlations

8.5.3 Influence of the Terrain

The influence of the terrain was investigated by calculating correlations for each valid sector of 10°, rather than using only one general correlation. Between the sectors, significant differences were found (Fig. 8.1). Two power curves were calculated using the correlations by sectors, or by taking the general correlation. Once again, significative differences were observed, varying between -10 and 10% (Fig. 8.2). However, in terms of the annual energy production (AEP), the variations were reduced to 0.65%. This is explained by a compensatory effect due to the fact that certain variations are positive and other negative. Overall, the use of a general correlation does not generate major errors over a long period, although involving significant differences between specific sectors.

8.5.4 Numerical Validation

The correlation is largely influenced by the cylindrical section near the root of the blade, and much less by the airfoils located at larger r/R. A common hypothesis is to use a cylinder drag coefficient $C_d = 1.2$. In reality, the C_d is function of the Reynolds number, and has to be calculated. Also, another correction has to be applied because of the finite length of the cylinder [3]. Another conclusion is that one of the major factor influencing the correlation comes from the acceleration of the flow produced by the attachment unit of the anemometer, as shown in Fig. 8.3a.

A U-beam has been added in the numerical simulation. The exact location and caracteristics of the fixation systems are not know precisely. Thus, two

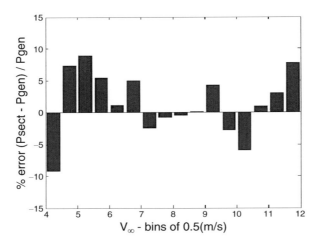

Fig. 8.2. Comparison between correlations by sectors and general correlation for all valid sectors on power curves

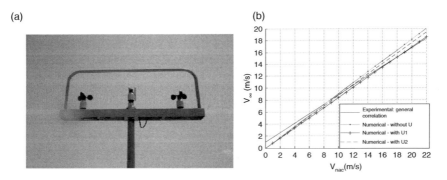

Fig. 8.3. (a) Attachment unit on the nacelle (b) Influence of the attachment unit

positions, U1 and U2, distant of 50 cm, have been analysed (Fig. 8.3b). The influence of the U-beam is significant, and can explain in part the differences observed between the numerical and experimental correlations.

8.6 Conclusion

Three main conclusions were drawn from experimental and numerical analysis. First, using the correlation provided by the manufacturer for any type of terrain can lead to important error. The second conclusion is that the use of a general correlation for all the valid sectors is not a bad practice. Finally, it has been numerically proven that for this WT, the attachment unit of the nacelle anemometer is an important cause of the flow perturbation.

References

1. R. Hunter et al. (2001): European Wind Turbine Testing Procedure Developments, Riso-R-1209
2. H. A. Madsen (1992): The Actuator Cylinder a Flow Model for Vertical Axis Wind Turbines, Aalborg University Centre, Denmark Thesis, Columbia University, New York
3. A. Smali, C. Masson (2005): On the Rotor Effects upon Nacelle Anemometry for Wind Turbine, Wind Engineering, in press

9

Pollutant Dispersion in Flow Around Bluff-Bodies Arrangement

Elżbieta Moryń-Kucharczyk and Renata Gnatowska

Summary. The aim of the paper is to discuss the relations between the complex velocity field around the bluff-bodies arrangement and the polluted gas concentration in that area. The main attention has been put on the role of oscillating component of velocity as a factor stimulating the pollutant diffusion process. The analysis has been performed for the 2D case of two sharp-edged bluff-bodies in tandem arrangement for which the strong level of oscillating component was available as the result of synchronization processes occurring in the flow surrounding the bodies. The mean concentration profiles of tracer gas (CO_2) for various inter-obstacle gap were measured in wind tunnel flow. The local characteristics of flow were obtained by the use of commercial CFD code (FLUENT). The discussion is partially based on the results of the previous data contained in works of Jarża & Gnatowska [1, 2], which revealed different flow regimes and critical body spacing at which dynamic properties of vortex structures around bluff-bodies system alter rapidly reflecting the stability mode change.

9.1 Introduction

The dispersion of pollutants in space around wind engineering structures, governed by convection and diffusion mechanism, depends strongly on the velocity field. To understand the phenomena related to the forming of concentration fields it is necessary to recognize the local features of the flow around the objects with the special emphasize for the mean velocity direction, random fluctuations, and periodical oscillations accompanying the vortex generation in bodies neighbourhood. The specific flow conditions generated around bluff-bodies arrangement make it possible to study the gas pollutant dispersion for the case of very complex velocity field typical for built environment. Curved streamlines, sharp velocity discontinuities, high level flow oscillations and non-homogenous turbulence disperse effluents in a complicated manner related to source configuration and object geometry.

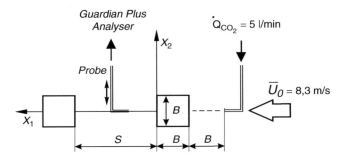

Fig. 9.1. Schematic presentation of the set-up and nomenclature

The sketch of the bluff-bodies arrangement considered in this paper (2D case) and the details of experiment performed in wind tunnel [1] by the use of carbon dioxide as tracer gas were shown in Fig. 9.1.

Such body configuration is known as a tandem. The inter-obstacle gap was changed in the range of nondimensional values $S/B = 0 \ldots 6$ (obstacle side dimension $B = 0.04$ m). The results of previous data, contained in works of Jarża & Gnatowska [1, 2], revealed different flow regimes and critical body spacing (S/B) at which drag coefficient and vortex shedding alter rapidly reflecting the stability mode change. Numerical simulations showed overall changes in flow pictures observed for increasing spacing ratio S/B. The program of this study consists of: measurement of the mean concentration profiles in the inter-body gap for different body spacing, comparison of concentration field with aerodynamic characteristics (obtained as a result of numerical simulation performed in ITM CzUT), and recognition of the role of vortex structure and related periodical oscillations.

9.2 Results of Measurements

The isolines of mean concentration for various inter-obstacle gap width $(S/B = 2, 4, 6)$, Fig. 9.2, are the first group of results. As one can see from the above figure the different distributions of CO_2 concentration are obtained in each case. In the case of $S/B = 2$ the maximum value of mean concentration replaces out of flow axis with an increase in the distance x_1/B. The region of higher concentration occurs distinctly outside the gap, overshooting down-stream cylinder. In accordance with [3] such behaviour could indicate the existence of one-body regime occurring for very small gaps between bodies. Maximum value of mean concentration is diverted from flow axis with an increase in the distance x_1/B in the case of $S/B = 6$ too, but for the distances $x_1/B > 1$ it moves to the centreline of the wake. Distinctly different picture of the isolines of mean concentration is observed for $S/B = 4$, what justifies the determination of this gap width as critical body spacing. In the above

9 Pollutant Dispersion in Flow Around Bluff-Bodies Arrangement   51

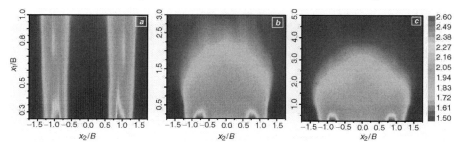

Fig. 9.2. Isolines of mean concentration in the inter-obstacle gap a) $S/B = 2$, b) $S/B = 4$, c) $S/B = 6$

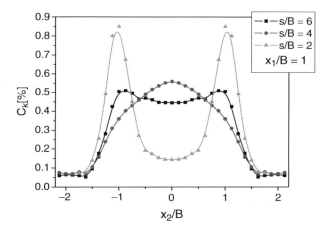

Fig. 9.3. Mean concentration profiles for various inter-obstacle gap width

case concentration peaks are distinctly moving towards the flow axis and for the distances $x_1/B > 0.75$ maximum values of mean concentration are in the centreline of the wake. The region of higher CO_2 concentration is noticeable in the whole inter-obstacle gap.

Mean concentration profiles for various inter-obstacle gaps and for the same position of $x_1/B = 1$ behind the upstream cylinder are presented in the Fig. 9.3. As one can notice, in the case of $S/B = 2$ and $S/B = 6$ maximum values of mean concentration are outside of the gap region, while for the gap width $S/B = 4$ maximum value of mean concentration is in the centreline of the wake. Computational profiles of the mean velocity components and numerical distribution of turbulence energy for the same cross section ($x_1/B = 1$) were used to find explanation. It was revealed that velocity distributions are similar so the question arises why the concentration profiles are different.

To find an explanation the results of computational modelling of velocity field performed by Gnatowska & Jarża [1] have been used. The numerical

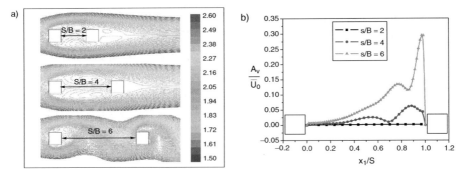

Fig. 9.4. Computational distributions of a) stream function b) amplitude of transverse velocity oscillations

simulation mentioned above provided the mean velocity distributions as well as characteristics of turbulence structure around tandem arrangement. The special emphasize has been put on intensity of velocity oscillation in space around bluff-bodies. The sample results, shown in Fig. 9.4a present the instantaneous computational distributions of stream function corresponding to different dynamic states of the flow in body system environment. The centreline distributions of amplitude of transverse velocity oscillations are also juxtaposed in Fig. 9.4b. One may suppose that transport of pollutants is intensified by periodical velocity component, generated in flow around bluff-bodies arrangement, especially for the critical body spacing. In this case the maximum of concentration, move forwards in the centreline of the wake.

9.3 Conclusions

In an experimental study of bluff-bodies tandem arrangement the significant changes have been observed in the concentration field of the tracer gas for different spacing ratio. Depending on body spacing the maximum concentration of CO_2 is localized outside of the gap (double peak) or migrates towards the centreline. The interbody gap is filled up by the emitted gas more intensively as the level of energy of oscillating velocity component increases (critical body spacing). The problem needs further studies.

References

1. Jarża J, Gnatowska R (2004) Lock-on effect on unsteady loading of rigid bluff - body in tandem arrangement. Proceedings of International Conference Urban Wind Engineering and Buildings Aerodynamics Cost C14. Rhode-St-Genese Belgium

2. Jarża A, Gnatowska R (2003) Simulation of lock-on phenomena in systems of bluff-bodies. Proceedings of 12th International Conference on Fluid Flow Technologies. Budapest Hungary
3. Havel B, Hangan H, Martinuzzi R (2001) Buffeting for 2D and 3D sharp-edged bluff-bodies. Journal of Wind Engineering and Industrial Aerodynamics 89:1369–1381

10

On the Atmospheric Flow Modelling over Complex Relief

Ivo Sládek, Karel Kozel and Zbyněk Jaňour

The paper deals with a mathematical and numerical investigation of a 3D-flow in the atmospheric boundary layer (ABL) over a complex topography situated around a Prague's agglomeration. The concept of a wall function is compared with a no-slip wall modelling. The mathematical model is based on full RANS equations considered in the conservative form. A simple algebraic turbulence closure is performed together with a stationary boundary conditions.

10.1 Mathematical Model

The atmospheric flow is supposed to be incompressible, viscous, turbulent and neutrally stratified and without any volume forces as well. The model is based on system of RANS equations hereafter re-casted in the conservative and vector form as follows:

$$\boldsymbol{F}_x + \boldsymbol{G}_y + \boldsymbol{H}_z = (K \cdot \boldsymbol{R})_x + (K \cdot \boldsymbol{S})_y + (K \cdot \boldsymbol{T})_z. \tag{10.1}$$

The system (10.1) is then modified according to the artificial compressibility method

$$\boldsymbol{W}_t + \boldsymbol{F}_x + \boldsymbol{G}_y + \boldsymbol{H}_z = (K \cdot \boldsymbol{R})_x + (K \cdot \boldsymbol{S})_y + (K \cdot \boldsymbol{T})_z, \tag{10.2}$$

where $\boldsymbol{W} = (p/\beta^2, u, v, w)^T$ stands for the vector of unknown variables, namely the pressure and the three velocity components, respectively. The vectors $\boldsymbol{F} = (u,\, u^2 + p/\varrho,\, uv, uw\,)^T$, $\boldsymbol{G} = (v,\, vu,\, v^2 + p/\varrho,\, vw\,)^T$, $\boldsymbol{H} = (w,\, wu,\, wv,\, w^2 + p/\varrho\,)^T$ denote the inviscid fluxes and $\boldsymbol{R} = (0,\, u_x,\, v_x,\, w_x\,)^T$, $\boldsymbol{S} = (0,\, u_y,\, v_y,\, w_y\,)^T$ and $\boldsymbol{T} = (0,\, u_z,\, v_z,\, w_z\,)^T$ represent the viscous ones. The density ϱ is supposed to be constant and $\beta \in \Re^+$ abbreviates the artificial sound speed. Finally, the parameter K refers to the turbulent diffusion coefficient, see (10.3) in Sect. 10.1.1.

10.1.1 Turbulence Model

A simple algebraic turbulence model is applied to the system (10.2) in order to close the problem. The turbulent diffusion coefficient K is expressed by

$$K = \nu + \nu_T, \quad \nu_T = l^2 \sqrt{\left(\frac{\partial u}{\partial z}\right)^2 + \left(\frac{\partial v}{\partial z}\right)^2}, \quad l = \frac{\kappa(z + z_0)}{1 + \kappa \frac{(z+z_0)}{l_\infty}} \qquad (10.3)$$

where ν_T, ν are the turbulent, laminar viscosities and l refers to the Blackadar's mixing length, $\kappa = 0.41$ is the von Karman constant, z_0 is the surface roughness length and l_∞ represents the mixing length for $z \to \infty$.

10.1.2 Boundary Conditions

The mathematical model (10.2) is equipped with the following stationary boundary conditions:

- Inlet: u-component is prescribed according to the power law with some exponent and $v = w = 0$.
- Wall-ground: the wall-function (10.4) or the no-slip condition ($u = v = w = 0$ at wall) are applied.
- Outlet, top face and side faces of computational domain: homogeneous Neumann conditions for all quantities are imposed.

The grid should be fine enough close to the wall to resolve all the gradients when the no-slip condition is applied. However, this increases the CPU-cost of such simulation due to a stability time-step limit for the explicit scheme that is actually used, see Sect. 10.1.3. On the other hand, the wall-function approach is less CPU-time consuming since the grid can be coarser at wall. The first inner grid cell is placed within the near-ground layer of a typical thickness about 50 m. The wall-function then reads

$$\sqrt{u^2 + v^2 + w^2} = \frac{u^*}{\kappa} \log\left(\frac{z_1 + z_0}{z_0}\right), \qquad (10.4)$$

where u^* denotes the friction velocity and z_1 refers to the distance of the center of the computational cell (where the wall-function is applied) from the wall.

10.1.3 Numerical Method

The mathematical model (10.2) is solved in the computational domain on a non-orthogonal structured grids made of hexahedral computational cells. It is expected to obtain a converged solution to the steady-state for all the unknowns and for the artificial time $t \to \infty$.

The finite volume method (cell centered) and a multi-stage explicit Runge-Kutta time integration scheme are applied to the system (10.2), see [1]. The scheme is theoretically second order accurate in space and time (on orthogonal grids) and it also needs to be stabilized by the artificial viscosity term of fourth order to remove a spurious oscillations from the flow-field generated by the use of central differences for space discretization.

10.2 Definition of the Computational Case

The computational domain is $43 \times 35 \times 1$ km long, wide and high and is discretized by:

1) no-slip wall modelling: $150 \times 100 \times 16$ mesh cells, horizontally uniform and exponentially distributed in the z-axis direction, $\Delta z_{\min} = 4.6$ m.
2) wall function modelling: $150 \times 100 \times 10$ mesh cells, horizontally uniform and exponentially distributed in the vertical direction, $\Delta z_{\min} = 28.3$ m.

The other parameters are: the mean free stream velocity $U = 10$ ms^{-1}, the characteristic wall-normal domain dimension $L = 1,000$ m and the corresponding Reynolds number $Re = U \times L/\nu = 6,7 \times 10^8$, the roughness parameter $z_0 = 1$ m and the power law exponent 0.3 and the friction velocity $u^* = 0.33$ ms^{-1} are used for the inlet velocity profile, see [2], [3].

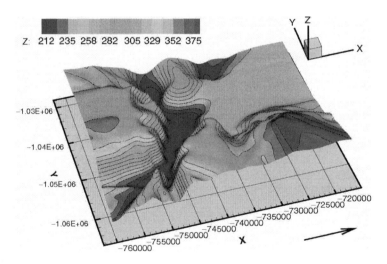

Fig. 10.1. The applied topography coloured by the geographical altitude (m), scale 1:1:17

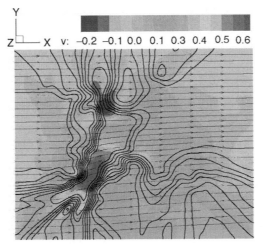

Fig. 10.2. Wall function case, v-colour-map

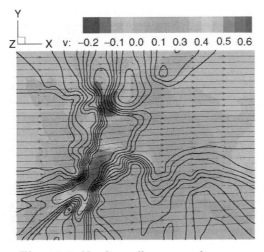

Fig. 10.3. No-slip wall case, v-colour-map

10.2.1 Some Numerical Results

The comparison of results obtained from both wall modelling approaches can be seen on the horizontal near-ground cutplanes coloured by the span-wise v-velocity component in (ms^{-1}) together with streamlines and also the geographical altitude contours, see Figs. 10.2 and 10.3.

The cutplanes are at the level of $\sim$14 m above the ground and one can see the overall agreement between the two figures and also the effect of the Prague's valley which decelerates the flow and deviates it from the x-axis.

10.3 Conclusion

A briefly mentioned mathematical-numerical model (10.2) has been applied to a real-case 3D ABL-flow problem over a complex topography. Also the two different wall-modelling approaches have been presented and some numerical results have been shown for comparison indicating an acceptable agreement. All the results have been obtained using a simple algebraic turbulence model.

Acknowledgment

The financial support for the present project is provided by the Grant 1ET 400760405 of GA AV ČR and by the Research Plan MSM No. 6840770003.

References

1. Sládek I., Kozel K., Janour Z., Gulíková E.: On the Mathematical and Numerical Investigation of the Atmospheric Boundary Layer Flow with Pollution Dispersion, In: COST Action C14 "Impact of Wind and Storm on City Life and Built Environment", pp. 233–242, ISBN 2-930389-11-7, VKI 2004
2. Bodnár T., Kozel K., Sládek I., Fraunie Ph.: Numerical Simulation of Complex Atmospheric Boundary Layer Flows Problems, In: ERCOFTAC bulliten No. 60: Geophysical and Environmental Turbulence Modeling, pp. 5–12, 2004
3. Beneš L., Bodnár T., Janour Z., Kozel K., Sládek I.: On the Complex Atmospheric Flow Modelling Including Pollution, In: "Wind Effects on Trees", pp. 183–188, ISBN 3-00-011922-1, Karlsruhe 2003

11

Comparison of Logarithmic Wind Profiles and Power Law Wind Profiles and their Applicability for Offshore Wind Profiles

Stefan Emeis and Matthias Türk

11.1 Wind Profile Laws

Two types of wind profile laws are often used for the vertical extrapolation of wind speeds $u(z)$ in the atmospheric surface-layer [1]: the theoretically derived logarithmic profile with stability corrections Ψ

$$u(z) = \left(\frac{u_*}{\kappa}\right) \left(\ln\left(\frac{z}{z_0}\right) - \Psi\left(\frac{z}{L_*}\right)\right) \tag{11.1}$$

with height z, roughness length z_0, Monin–Obukhov length L_*, friction velocity u_*, and von Kármán's constant $\kappa = 0.4$ and the empirical power law

$$u(z) = u(z_A) \left(\frac{z}{z_A}\right)^n \tag{11.2}$$

with reference height z_A and Hellmann exponent n. Due to its mathematical simplicity (11.2) is widely used for wind energy purposes. The first part of this study (Chap. 11.2) investigates the possible approximation of (11.1) by (11.2), the second part (Chap. 11.3) compares (11.1) and (11.2) to measured offshore wind profiles.

11.2 Comparison of Profile Laws

In extension to existing studies (e.g. [2]) the slope and the curvature of the logarithmic and the power law profiles should coincide in a selected height simultaneously in order to give a good fit over a wider height range. The surface roughness and thermal stratification conditions for which such a simultaneous coincidence is possible are calculated analytically (see the full equations in [3]). It turns out that for neutral and unstable conditions slopes and curvatures of the two profiles (11.1) and (11.2) cannot coincide simultaneously. Only an approximate coincidence is found in the limit $z_0 \to 0$ for very smooth surfaces.

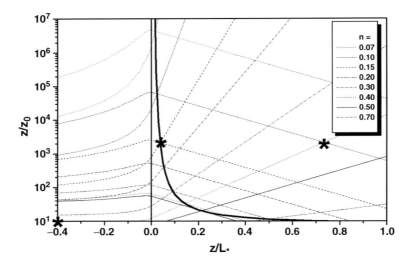

Fig. 11.1. Diagram showing the solution of (11.3) (*bold line*). Roughness decreases along the y-axis and stable stratification increases along the x-axis. Thin lines towards the upper right indicate power law and logarithmic profiles with equal slope, lines towards the lower right those profiles with equal curvature. Asterisks mark positions for Fig. 11.2

A perfect coincidence is possible in stably stratified flow, if the following relation between roughness and stratification holds (see Fig. 11.1):

$$\ln\left(\frac{z}{z_0}\right) = 2 + \frac{1}{4.7(z/L_*)}. \tag{11.3}$$

The practical result is that (11.2) offers a nearly perfect fit to (11.1) under stable conditions for certain surface roughness conditions and a good approximation under neutral and unstable conditions in the limit of very smooth surfaces.

11.3 Application to Offshore Wind Profiles

Offshore wind and turbulence profiles from the 100 m mast on the FINO1 platform in the German Bight 45 km off the coast are currently processed at our lab (project OWID, funded by the German Ministry of the Environment, BMU by grant no. 0329961). Measurement heights are 33.5, 41, 51, 61, 71, 81, 91, and 102.5 m. For comparison with (11.1) and (11.2) wind data from 2004 have been used. According to the difference between air temperature in 41 m height and water temperature the profiles have been lumped into several stability classes. Figure 11.3 shows exemplary annual mean profiles for very

11 Comparison of Logarithmic and Power Law Wind Profiles

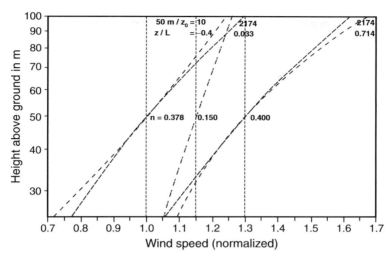

Fig. 11.2. Wind profiles from (11.1) (*dotted*) and from (11.2) (*dashed*) for the three positions marked in Fig. 11.1. Wind speeds are normalized to wind speeds in 50 m height. The neutral profiles have been shifted 0.15 to the right and the stable profiles have been shifted 0.3 to the right for clarity

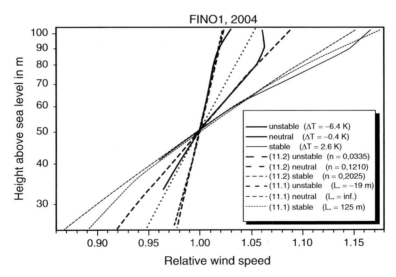

Fig. 11.3. Comparison of measured wind profiles with wind profile laws (11.1) and (11.2) (normalized to speeds in 51 m height) for three different stability classes

unstable ($-6.9\,\mathrm{K} < \Delta T < -5.9\,\mathrm{K}$), neutral ($-0.9\,\mathrm{K} < \Delta T < 0.1\,\mathrm{K}$), and very stable ($2.1\,\mathrm{K} < \Delta T < 3.1\,\mathrm{K}$) stratification. These profile are approximated by (11.1) and (11.2) in such a way that the measurements in 51 m and 61 m height are matched. For the logarithmic profiles a roughness length of 0.0001 m has been assumed. For neutral stratification no fit of the logarithmic profile to the measured data was possible because this profile is fixed through the inevitable choice $L_* = \infty$ for the Monin–Obukhov length.

The fits are satisfying over the whole height range only for stable stratification. The logarithmic profile from (11.1) is closer to the measurements than the power law profile from (11.2). For neutral conditions and unstable conditions below 50 m the fits are not so well. The measured profile for neutral conditions below 81 m is more slanted to the right than the truly neutral logarithmic profile indicating that we are already in the upper part of the surface layer. Above 81 m the measured wind speeds under neutral and stable conditions increase slower with height. Obviously, we reach here already the Ekman layer. Under unstable stratification the surface layer is higher and thus no drop in the wind speed increase with height can be observed below 102.5 m.

11.4 Conclusions

Within the height range of their validity (i.e. the surface layer of the atmospheric boundary layer) a power law wind profile can only approximate a logarithmic wind profile over a wider height range in case of stable thermal stratification for certain surface roughnesses (see (11.3)). For neutral and unstable stratification this approximation is possible only for very smooth surfaces.

There is an indication that for neutral and stable thermal stratification of the air offshore wind profiles in heights above about 80 m above the sea are already above the surface layer within which (11.1) is valid. Nevertheless a description of offshore profiles by (11.1) and (11.2) seem to be possible at least under stable conditions. But it remains open whether the Monin–Obukhov length L_* used here to fit the logarithmic profile in 51 to 61 m height to the measured profile resembles to L_* in the true surface layer in the first ten or twenty meters above the sea surface.

References

1. Stull RB (1988) An Introduction to Boundary Layer Meteorology. Kluwer, Dordrecht, pp. 666
2. Sedefian L (1980) On the vertical extrapolation of mean wind power density. J Appl Meteorol 19:488–493
3. Emeis S (2005) How Well Does a Power Law Fit to a Diabatic Boundary-Layer Wind Profile? DEWI Magazine 26:59–62

12

Turbulence Modelling and Numerical Flow Simulation of Turbulent Flows

Claus Wagner

12.1 Summary

Some popular techniques used in simulations of turbulent flows are presented and discussed. It is shown how the (k, ω) turbulence model and two different dynamic subgrid scale models perform in Reynolds-averaged Navier–Stokes simulations (RANS) and Large-Eddy Simulations (LES) of turbulent channel flow, respectively. Besides, some drawbacks of the eddy viscosity concept which is the basis for most turbulence models are discussed.

12.2 Introduction

Phenomenologically most flows can be categorized into the laminar and the turbulent flow regime. Any disturbance, which is damped by molecular dissipation in a laminar flow, grows to form a turbulent, chaotic-like, time-dependent, three-dimensional flow field, if the relative impact of the molecular dissipation is reduced. The latter is expressed by an increase of the Reynolds number $Re = u/(\nu l)$ (ν represent the kinematic viscosity, l and u the length and velocity scales of the flow). Turbulent flows are characterized by fluctuations on a wide range of scales the size of which decrease with increasing Reynolds numbers. In order to properly resolve all scales in a so-called direct numerical simulation (DNS) one has to specify extremely fine grids, resulting in unaffordable computing times for high Reynolds number flows. To overcome this, researchers developed turbulence models, which approximate the effect of smaller scales. This is necessary since most environmental or technical relevant turbulent flows are characterized by very high Reynolds numbers. Hence, turbulence modelling is one of the key problems in numerical flow simulation.

66   C. Wagner

12.3 Governing Equations

In (12.1) the dimensionless incompressible Navier–Stokes equations which contain the Reynolds number Re are presented in cartesian coordinates using Einstein's summation convention

$$\frac{\partial u_i}{\partial x_i} = 0, \quad \frac{\partial u_i}{\partial t} + \frac{\partial u_i u_j}{\partial x_j} = -\frac{\partial p}{\partial x_i} + \frac{1}{Re}\frac{\partial}{\partial x_j}\underbrace{\left(\frac{\partial u_i}{\partial x_j} + \frac{\partial u_j}{\partial x_i}\right)}_{2S_{ij}}. \qquad (12.1)$$

Almost any known analytical solution of (12.1) is valid only in the laminar regime, i.e. for low Reynolds numbers. For higher Reynolds numbers, disturbances which are introduced for example by imperfect or rough walls tend to grow into three-dimensional vortical structures. Due to stretching and tilting of the associated vortex lines an initially simple flow pattern changes into complicated turbulence, which is characterized by irregularity in space and time, increased dissipation, mixing and nonlinearity.

Often, one is not interested in all details of the time-dependent three-dimensional velocity and pressure field, but in the behaviour of the statistical mean or of the large scales. To reduce the level of detail of the solution any turbulent variable f is split into a filtered value $\langle f \rangle$ and its fluctuation $f - \langle f \rangle$.

The corresponding filtering of (12.1) leads to the filtered Navier–Stokes equations which contains the extra nonlinear term τ_{ij} in comparison to (12.1).

$$\frac{\partial \langle u_i \rangle}{\partial x_i} = 0, \quad \frac{\partial \langle u_i \rangle}{\partial t} + \frac{\partial \langle u_i \rangle \langle u_j \rangle}{\partial x_j} + \frac{\partial \tau_{ij}}{\partial x_j} = -\frac{\partial \langle p \rangle}{\partial x_i} + \frac{1}{Re}\frac{\partial}{\partial x_j}\underbrace{\left(\frac{\partial \langle u_i \rangle}{\partial x_j} + \frac{\partial \langle u_j \rangle}{\partial x_i}\right)}_{2\langle S_{ij} \rangle}.$$

$$(12.2)$$

Using statistical averaging as a filter function one obtains the so-called Reynolds stress tensor $\tau_{ij} = \langle u_i'' u_j'' \rangle$, where u_i'' and u_j'' denote the statistical velocity fluctuations. Then (12.2) represents the RANS equations. The statistical approach is associated with the highest loss of information and with a closure problem which is not satisfactorily solved. Spectral information is completely lost, since any statistical quantity is an average over all turbulent scales. The obtained flow field describes the mean flow, which is enough for many applied problems, while the turbulent information is described with the Reynolds stress tensor.

The LES technique resembles a compromise between RANS and DNS since it allows to predict the dynamics of the large turbulent scales while the effect of the fine scales are modeled with a subgrid-scale model. The governing equations which have to be solved in a LES are also derived applying a filter function on the Navier–Stokes equations (12.1) to formally remove scales with a wavelength smaller than the grid mesh. To distinguish top-hat filtering from statistical averaging in (12.2) $\langle u_i \rangle$ and $\langle p \rangle$ are substituted by $\overline{u_i}$ and $\overline{p}$ and the subgrid scale tensor $\tau_{ij} = u_i' u_j'$ where u_i' denotes the subgrid scale velocity fluctuations.

12.4 Direct Numerical Simulation

Reliably solving (12.1) without any additional turbulence and/or transition model in a DNS requires accurate numerical methods, which do not inhibit significant artificial viscosity. Spectral methods, which provide very accurate spatial discretization have been used successfully in the past mainly for simple geometries. Besides spectral methods, finite difference methods based on second- or fourth-order accurate central difference schemes are widely used for DNS of turbulent flows. They are in general easily applicable in complex computational domains. Further, as pointed out by Choi et al. [1] it is possible to obtain "spectral" accuracy if the first and second order moments of the velocity fluctuations are considered for the same number of grid points.

12.5 Statistical Turbulence Modelling

The unknown Reynolds stress tensor $\langle u_i'' u_j'' \rangle$ in (12.2) is a symmetric tensor, which, for statistically three-dimensional flows, contains six different elements. In most applied flow problems $\langle u_i'' u_j'' \rangle$ is approximated using one- or two equation turbulence models. One popular two equation turbulence model is the so-called (k, ω)-model by Wilcox [9] (ω does not correspond to the vorticity). This model which allows to predict seperated flows better than the well-known (k, ω)-model reads:

$$-\langle u_i'' u_j'' \rangle = \nu_t 2 \langle S_{ij} \rangle - \frac{2}{3} k \delta_{ij}, \qquad \nu_t = \frac{k}{\omega}, \qquad k = \frac{\langle u_i'' u_i'' \rangle}{2}, \qquad \omega = \frac{\varepsilon}{c_\mu k} \tag{12.3}$$

with the unknown properties k and ω. Closure is obtained solving the following transport equation for these two unknowns and using a set of empirical constants $(c_\mu, \sigma_k, c_{\omega 1}, c_{\omega 2}, \sigma_\omega)$.

$$\langle u_i \rangle \frac{\partial k}{\partial x_i} = c_\mu \frac{k^2}{\varepsilon} 2 \langle S_{ij} \rangle \langle S_{ij} \rangle + \frac{1}{Re_\tau} \frac{\partial (1 + \frac{\nu_t}{\sigma_k \nu}) \frac{\partial k}{\partial x_i}}{\partial x_i} - \varepsilon, \tag{12.4}$$

$$\langle u_i \rangle \frac{\partial \omega}{\partial x_i} = c_{\omega 1} 2 \langle S_{ij} \rangle \langle S_{ij} \rangle - c_{\omega 2} \omega^2 + \frac{1}{Re_\tau} \frac{\partial (1 + \frac{\nu_t}{\sigma_\omega \nu}) \frac{\partial \omega}{\partial x_i}}{\partial x_i}. \tag{12.5}$$

In Fig. 12.1 the mean axial velocity component and the mean pressure computed by Frahnert and Dallmann [2] in a RANS calculation of the fully developed turbulent channel flow with the (k, ω)-model are compared to the DNS results of Kim et al. [4] for a Reynolds number $Re_\tau = u_\tau H / \nu = 360$ (u_τ denotes the friction velocity and H the channel height). While the profiles of the mean axial velocity component $\langle u \rangle$ in Fig. 12.1 compare very well, the mean pressure distributions in Fig. 12.1b reveal strong discrepancies. Considering the comparison of the turbulence intensities in Fig. 12.2, which are the four different components of the Reynolds stress tensor, good agreement is obtained for the Reynolds stress $\langle u' v' \rangle$, but the normal stresses $\langle u'^2 \rangle$, $\langle v'^2 \rangle$ and $\langle w'^2 \rangle$ compare rather poorly.

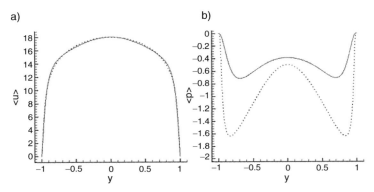

Fig. 12.1. Mean axial velocity components (**a**) and mean pressure (**b**) of a fully developed turbulent channel flow. *Solid lines*: DNS by Kim et al., *dotted lines*: RANS by Frahnert and Dallmann

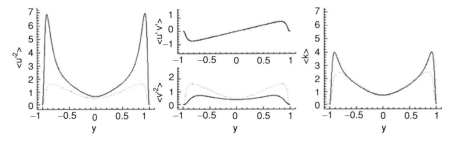

Fig. 12.2. Components of the Reynolds stress tensor of a fully developed turbulent channel flow. *Solid lines*: DNS by Kim et al., *dotted lines*: RANS by Frahnert and Dallmann

12.6 Subgrid Scale Turbulence Modelling

12.6.1 Eddy Viscosity Models

Using the eddy viscosity concept to approximate τ_{ij} in an LES was first proposed by the meteorologist Smagorinsky, who simulated large scale atmospheric motions. In Smagorinsky's model [7] the eddy viscosity ν_t is proportional to the grid length scale $\overline{\Delta}$, the local strain rate $|\overline{S}| = 1/2\sqrt{\overline{S}_{ij}\overline{S}_{ij}}$ and the constant C_s.

$$\tau_{ij} = \overline{u'_i u'_j} = \underbrace{(-C_s\overline{\Delta})^2 \cdot |\overline{S}|}_{\nu_t} \cdot \overline{S}_{ij} + \frac{1}{3}\overline{u'_k u'_k}\delta_{ij}. \tag{12.6}$$

Using the dynamic process by Germano [3] the unknown constant C_s can be estimated assuming that its value does not change between two differently filtered flow fields. Therefore, in addition to the grid filter $\overline{\Delta}$, Germano proposed to use a second filtering level $\widetilde{\overline{\Delta}} > \overline{\Delta}$. Then the subgrid scale stress tensors on the two filtering levels read

$$\tau_{ij} = \overline{u_i u_j} - \overline{u}_i \overline{u}_j, \quad T_{ij} = \widetilde{\overline{u_i u_j}} - \widetilde{\overline{u}}_i \widetilde{\overline{u}}_j. \tag{12.7}$$

For both filtering levels one applies the Smagorinsky models

$$\tau_{ij} - \frac{1}{3}\delta_{ij}\tau_{kk} = 2C_s\overline{\Delta}^2 \mid \overline{S} \mid \overline{S}_{ij}, \qquad T_{ij} - \frac{1}{3}\delta_{ij}T_{kk} = 2C_s\widetilde{\overline{\Delta}}^2 \mid \widetilde{\overline{S}} \mid \widetilde{\overline{S}}_{ij}. \tag{12.8}$$

Similar to the subgrid scale stress tensor τ_{ij}, the Leonard stress tensor

$$L_{ij} = \widetilde{\overline{u}_i \overline{u}_j} - \widetilde{\overline{u}}_i \widetilde{\overline{u}}_j = T_{ij} - \widetilde{\tau}_{ij} \tag{12.9}$$

is unknown in a simulation. But replacing T_{ij} and τ_{ij} by an approximation like the Smagorinsky model of (12.6) leads to six equations for the unknown Smagorinsky constant C_s.

$$L_{ij} = -2C_s\underbrace{(\widetilde{\overline{\Delta}}^2 \mid \widetilde{\overline{S}} \mid \widetilde{\overline{S}}_{ij} - \overline{\Delta}^2 \widetilde{\mid \overline{S} \mid \overline{S}_{ij}})}_{M_{ij}}. \tag{12.10}$$

To obtain a single constant C_s Lilly [5] suggested to minimize the error tensor ε_{ij} applying the least square formulation in the sense

$$\varepsilon_{ij} = L_{ij} + 2C_s M_{ij} \quad \Longrightarrow \quad C_s = -\frac{1}{2}\frac{L_{ij}M_{ij}}{M_{ij}M_{ij}}. \tag{12.11}$$

Unfortunately solving (12.11) results in a C_s field which strongly varies in space and in time and with a significant number of negative values. In many simulations large negative values tend to destabilize the numerical simulation since they lead to a nonphysical growth of the resolved scale energy. To overcome this, a statistical averaged $\langle C_s \rangle$ has been used for almost all reported simulations.

12.6.2 Scale Similarity Modelling

Applying filtering with two filter widths Liu et al. [6] introduced the unknown constant C_L in a scale similarity model $\tau_{ij} = C_L L_{ij}$. Later Wagner [8] proposed to modify the dynamic process to determine C_L. In his approach τ_{ij} and L_{ij} in (12.9) are substituted by (12.8). Just as in Germano's dynamic model he obtains a set of equations for the unknown constant C_L

$$\overline{\Delta}^2 \mid \overline{S} \mid \overline{S}_{ij} = C_L(\widetilde{\overline{\Delta}}^2 \mid \widetilde{\overline{S}} \mid \widetilde{\overline{S}}_{ij} - [\overline{\Delta}^2 \widetilde{\mid \overline{S} \mid \overline{S}_{ij}}]) \tag{12.12}$$

which can be solved applying the least square method to minimize the error. Applying his model in various LES of turbulent channel flow for different

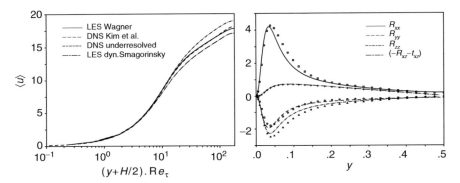

Fig. 12.3. Mean velocity profiles calculated in LES and DNS (*left*). Profiles of the deviatoric part of the Reynolds stress tensor (*right*) calculated in LES with the scale similarity model and compared to DNS data by Kim et al.) (denoted by symbols)

Reynolds numbers he obtained values for C_L of the order of 1, but with a significant variation in wall normal direction. Again C_L was statistically averaged to enforce the necessary smoothness of C_L, which is necessary for conserving the high correlation between τ_{ij} and L_{ij}

In order to give an impression on the performance of the different described subgrid scale models, results of LES of a fully developed turbulent channel flow performed for a Reynolds number $Re_\tau = 360$ are presented in Fig. 12.3. For both, the mean velocity profile and the resulting deviatoric part of the Reynolds stress tensor $R_{ij} = \langle u_i u_j \rangle - \langle u_i \rangle \langle u_j \rangle - 1/3(\langle u_k u_k \rangle - \langle u_k \rangle \langle u_k \rangle)\delta_{ij}$ which are compared to the according data of Kim et al. excellent agreement is obtained while the mean values of the underresolved DNS and of the LES with the Smagorinky model shows considerable discrepiances.

12.7 Conclusion

The analysis of the eddy viscosity concept which is applied in most RANS and traditional LES turbulence models showed that such models are not able to reliably predict nonisotropic flows. Therefore, if such simulations are performed to optimize or analyse rotor blades of wind turbines, the obtained results have to be interpreted with caution.

References

1. Choi H, Moin P, Kim J (1992) Turbulent drag reduction: studies of feeedback control and flow over riplets, Rep. TF-55, Department of Mechanical Engineering, Stanford University, Stanford, CA

2. Frahnert H, Dallmann U Ch (2002) Examination of the eddy-viscosity concept regarding its physical justification. In: Wagner S, Rist U, Heinemann J, Hilbig R. (eds) New Results in Numerical and Experimental Fluid Mechanics III Springer, Berlin Heidelberg New York
3. Germano, M: Turbulence: the filtering approach; J. Fluid Mech. A 5:1282-84 (1992)
4. Kim J, Moin P, Moser R (1987) J. Fluid Mech. A 177:133–166
5. Lilly DK (1992) Phys. Fluids A 4:633–635
6. Liu S, Meneveau D, Katz J (1994) J. Fluid Mech. 275:83–119
7. Smagorinsky J (1963) Mon. Weather Rev. 91–99
8. Wagner C (2001) Z. Angew. Math. Mech. 81, Suppl. 3:491–492
9. Wilcox D C (1988) AIAA Journal, 26:1299–1310

13

Gusts in Intermittent Wind Turbulence and the Dynamics of their Recurrent Times

François G. Schmitt

Summary. Regions interesting for the installation of windmills possess large mean wind velocity, but this is also associated to large Reynolds numbers and huge velocity fluctuations. Such fully developed turbulence is generally characterized by intermittent fluctuations on a wide range of scales. There are many gusts (large fluctuations of the wind velocity at small scales), producing extreme loading on wind turbine structure. Here we consider the dynamics of gusts: the recurrent times of large fluctuations for a fixed time increment, and also "inverse times" corresponding to the times needed to have wind fluctuations larger than a fixed threshold. We discuss theoretically these two different quantities, and provide an example of data analysis using wind data recorded at 10 Hz. The objective here is to provide theoretical relations that will help for the interpretation of available data, and for their extrapolation.

13.1 Introduction

Wind turbines need theoretically large and constant wind velocities to be optimally efficient. However, this is not possible since large velocities correspond to high Reynolds numbers and fully developed turbulence situations, characterized by intermittent fluctuations of the wind velocity having highly nonlinear and non-Gaussian properties [1–3]. Turbulent wind fluctuate at all scales in intensity and in direction, producing high loading on wind turbine structure, large fluctuations on wind turbine force and moments, and high variability on wind turbine power output. Here we focus on small-scale wind fluctuations called wind gusts, which produce most of the fatigue loading on turbine structures; we consider theoretically, with data analysis applications, the dynamics of these gusts: the recurrent times of large fluctuations for a fixed small time increment (also called "return times" or "waiting times"), and also "inverse times" corresponding to the times needed to have wind fluctuations larger than a fixed threshold.

74 François G. Schmitt

13.2 Scaling and Intermittency of Velocity Fluctuations

The scaling and intermittent properties of wind velocity have been studied in the turbulence community for several decades (see reviews in [1,3,4]). We recall here some basic properties of this framework. The Richardson–Kolmogorov energy cascade develops on the inertial range, between the large injection scale (of the order of several tens or hundreds of meters) towards dissipative scales (of the order of millimeters). In the inertial range, wind velocity fluctuations are scaling: they possess a power-law spectrum

$$E(k) \simeq k^{-\beta} \tag{13.1}$$

with $\beta = 5/3$ for K41 turbulence [5]; intermittency however leads to a β value slightly larger than 5/3. Here we may define intermittency as the property of having large fluctuations at all scales, having a correlated structure: large fluctuations are much more frequent than what would be obtained for Gaussian processes [1, 3, 4]. Intermittency is then classically characterized considering the probability density function (pdf) or the moments of the velocity fluctuations $\Delta V_\ell = |V(x + \ell) - V(x)|$. For large mean velocities, Taylor's hypothesis may be invoked to relate time variations $\Delta V_\tau = |V(t+\tau) - V(t)|$ to spatial fluctuations; one may then study time intermittency of wind velocity considering the scaling property of moments of order $q > 0$ of the fluctuations ΔV_τ

$$\langle (\Delta V_\tau)^q \rangle \simeq \tau^{\zeta(q)}, \tag{13.2}$$

where $\zeta(q)$ is the scale invariant moment function, which seems rather universal for fully developed turbulent flows in the laboratory or the atmosphere, for moments small than 7. This function is nonlinear and concave; $\zeta(3) = 1$ is a fixed point; $\zeta(2) = \beta - 1$ relates the second order moment to the power spectrum scaling exponent; the knowledge of the full $(q, \zeta(q))$ curve for integer and noninteger moments provides a full characterization of velocity fluctuations at all scales and all intensities.

13.3 Gusts for Fixed Time Increments and Their Recurrent Times

Wind gusts may be generally defined choosing a time increment τ and a threshold velocity δ: a gust corresponds to a situation when $|V(t_1) - V(t_2)| > \delta$ with $|t_1 - t_2| < \tau$ [6]. In practice, for such event to occur, the condition $|V(t_1) - V(t_2)| > \delta$ should also be realized for some times verifying $|t_1 - t_2| = \tau$. We then choose here to fix a small time increment τ and to consider large fluctuations at this scale. In the following we choose $\tau = 3$ since $3 - s$ extreme gust correspond to a standard of the American Society of Civil

Engineers [7]. However this scale belongs to the inertial range; it is usually not a characteristic time scale, so that the results obtained here are expected to be valid for any other scale τ belonging to the inertial range.

Here we consider the recurrent times of gust events: the times between successive gust events. The recurrent times of meteorological events have practical importance, and correspond to the knowledge of some dynamical properties of wind fluctuations. Here we not only focus on the mean return times, but also on their pdf. We consider the evolution of this pdf with increasing velocity thresholds δ. For illustration purposes, we have taken an atmospheric turbulent velocity data base: velocity measurements taken 25 m from the ground in south-west France, with a sonic velocimeter sampling at 10 Hz (see [8, 9]). We analyze here three portions of duration 55 min each, recorded under near-neutral stability conditions. A portion of the data is shown in Fig. 13.1: this illustrates the intermittent behavior of wind turbulence. The power spectrum of the 3 portions is shown in Fig. 13.2, in log–log plot, indicating a very nice $-5/3$ scaling law over most of the available dynamics (between $0.2s$ and about 10 min). We have then transformed the velocity time series into an amplitude increment time series: $Y_3(t) = |V(t + \tau_0) - V(t)|$ with $\tau_0 = 3s$. The new time series is shown in Fig. 13.3, together with an example of threshold ($\delta = 1$ m^{-1}), and the associated recurrent times. Successive return time form a new time series which is represented in Fig. 13.4.

An interesting question is now to evaluate the tail behavior of the pdf of return times: since gust events are associated to large return times, their probability of occurrence is given by the tails. There are in fact many results providing the form of the tail of the pdf of recurrent times (also called first-passage times) for scaling processes. One of the most classical scaling

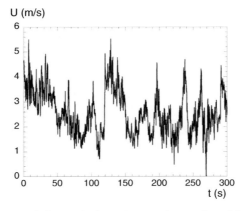

Fig. 13.1. A portion of the turbulent time series analyzed here for illustration purposes

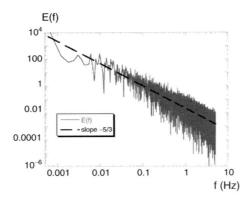

Fig. 13.2. The power spectrum of atmospheric turbulent data, in log–log plot, with a straight *dotted line* of slope $-5/3$ for comparison

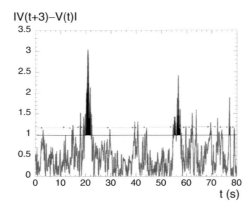

Fig. 13.3. A portion of the $3-s$ increment time series; large fluctuations correspond to $3-s$ gusts. Gusts associated to a velocity $> 1\,\mathrm{m\,s}^{-1}$ are shown, together with the construction of their recurrent times

process is Brownian motion: using dimensional arguments, Feller [10] showed that the pdf of their first-passage times is of the form:

$$p(T) \simeq T^{-\mu} \tag{13.3}$$

with $\mu = 3/2$. For fractional Brownian motion characterized by an exponent $H = \zeta(1)$, it has been shown that the form of (13.3) is still valid, with $\mu = 2 - H$ [11, 12]. For multifractal processes, the choice of a given threshold δ selects a fractal black-and-white process: the "gust" state occurs on a set whose support has a fractal dimension d_s; the larger the threshold δ, the smaller the dimension d_s characterizing the occurrence of gust events. In such case, the tail behavior of recurrence times has been shown to obey (13.3), with

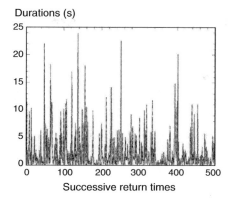

Fig. 13.4. A series of 500 consecutive return times for the increment $\tau = 3s$ and the threshold $\delta = 1\,\mathrm{m\,s^{-1}}$

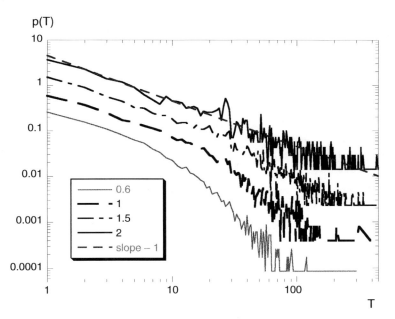

Fig. 13.5. The pdf of recurrent times estimated for various velocity thresholds (0.6, 1, 1.5 and $2\,\mathrm{m\,s^{-1}}$) respectively, from bottom to top. A straight line of slope -1 is shown for comparison with the $2\,\mathrm{m\,s^{-1}}$ threshold curve

$\mu = d_s + 1$ [13]. Return times are thus very intermittent, with a hyperbolic pdf tail associated to divergence of moments of order $\mu - 1$.

This has been tested on experimental data: Figure 13.5 shows four different pdfs in log–log plot, estimated for increasing thresholds from 0.6 to $2\,\mathrm{m\,s^{-1}}$: the hyperbolic tail is visible, with a slope decreasing with increasing thresholds, confirming some previously observed results [14]. As already noticed in [13] in the context of rainfall, such tail behavior indicates that mean return times would theoretically diverge in the limit of an infinite number of realizations (since $\mu < 2$); the estimated mean return time depends in fact on the number of realizations taken for the statistics. Let us note also that the largest gusts correspond to a dimension d_s close to 0, and thus a pdf of tail close to -1: this is confirmed in Fig. 13.5, for the largest threshold of 2 m s^{-1}: a straight line of slope -1 is shown for comparison. Return times for such extreme gusts correspond to sporadic processes, whose probability density is not normalizable ($\int p(t)\mathrm{d}t$ diverges) [15, 16].

13.4 The Dynamics of Inverse Times: Times Needed for Fluctuations Larger than a Fixed Velocity Threshold

We consider here another dynamical property of wind velocity fluctuations, that could also be interesting in the framework of wind energy applications: the "inverse times," i.e., successive first times necessary to observe a given velocity difference. This has been called "inverted structure functions" by Jensen [17] who introduced the idea, and is now known as "inverse structure functions" [18–20]. This name is justified by the fact that usual structure functions correspond to wind velocity increments statistics for given time increments; inverse structure functions correspond to time statistics for given velocity increments. Practically, these inverse times may be estimated the following way: if t_i is the time associated to a given inverse time, the next inverse time is $T_i = t_{i+1} - t_i$ with $t_{i+1} = \min\left(t \geq t_i; |V(t) - V(t_i)| \geq \delta\right)$

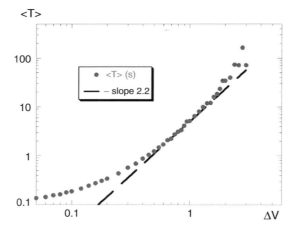

Fig. 13.6. Mean inverse times, estimated for various velocity thresholds. A straight line of slope 2.2 is shown for comparison

(the initial t_1 is the first time of the time series). We have estimated such inverse times time series, for different velocity thresholds from 0.02 to $3\,\mathrm{m\,s}^{-1}$, and reported in Fig. 13.6 the mean inverse times. This shows that the mean inverse times increase with the threshold approximately as a power-law for large thresholds, in agreement with previous results [17–20].

References

1. Frisch U (1995) Turbulence; the legacy of A. N. Kolmogorov. Cambridge University Press, Cambridge
2. Pope S (2000) Turbulent flows. Cambridge University Press, Cambridge
3. Vulpiani A, Livi R (eds) (2004) The Kolmogorov legacy in physics. Springer, Berlin Heidelberg New York
4. Schertzer D, Lovejoy S, Schmitt F, Chigirinskaya Y, Marsan D (1997) Fractals 5:427–471
5. Kolmogorov A N (1941) CR Acad. Sci. URSS 30:301–305
6. Mann J (2000) draft paper
7. Cheng E (2002) J. Wind Eng. Ind. Aerodyn. 90:1657–1669
8. Schmitt F, Schertzer D, Lovejoy S, Brunet Y (1994) Nonlin. Proc. Geophys. 1:95–104
9. Schmitt F, Schertzer D, Lovejoy S, Brunet Y (1996) Europhys. Lett. 34:195–200
10. Feller W (1971) An introduction to probability theory and its applications, vol. II. Wiley, New York
11. Hansen A, Engoy T, Maloy K J (1994) Fractals 2:527–533
12. Ding M, Yang W (1995) Phys. Rev. E 52:207–213
13. Schmitt F G, Nicolis C (2002) Fractals 10:285–290
14. Boettcher F, Renner C, Waldl H P, Peinke J (2003) Bound. Lay. Meteor. 108:163–173
15. Gaspard P, Wang X J (1988) Proc. Natl. Acad. Sci. USA 85:4591–4595
16. Wang X J (1992) Phys. Rev. A 45:8407–8417
17. Jensen M H (1999) Phys. Rev. Lett. 83:76–79
18. Beaulac S, Mydlarski L (2004) Phys. Fluids 16:2126–2129
19. Pearson B R, van de Water W (2005) Phys. Rev. E 71:036303
20. Schmitt F G (2005) Phys. Lett. A 342:448–458

14

Report on the Research Project OWID – Offshore Wind Design Parameter

T. Neumann, S. Emeis, and C. Illig

14.1 Summary

The chapter gives an overview of the research project OWID that has been launched in mid-2005. Aim of the project is to make proposals to improve offshore related standards and guidelines on the basis of measured FINO1 data and CFD calculations. Some examples for the motivation of the research project and a first glance on some preliminary results are given.

The FINO1 platform [1, 2], which is installed in 2003 about 45 km off the island Borkum, is equipped with a met mast with a height of about 100 m and records the long-term meteorological and oceanographic conditions in the North Sea.

Within the project OWID the FINO-data are used to reduce incomplete knowledge when adapting wind turbines to the maritime conditions. We start with a thorough evaluation of the acquired FINO1 data with the focus on the mechanical loads a future wind turbine is exposed to. In addition to the undisturbed wind field the disturbed wind stream within the wake field is simulated by CFD models as we think that the major part of the load originates in the wake fields. Both undisturbed and disturbed wind fields are used to calculate the loads on a realistic offshore wind turbine with regard to the lay out and the life time.

14.2 Relevant Standards and Guidelines

At present external wind conditions in the offshore regime are defined in guidelines by GL, IEC and DNV [3, 5]. It is an usual approach within the guidelines to define a reference and average wind speed for certain classes together with parameters for different turbulence regimes as shown in Fig. 14.1 for the GL-offshore guideline [5]. While the DNV [3] and IEC [4] guidelines still refer to the onshore related IEC-61400-1 [6], the GL-Offshore [5] already introduced a special subclass C with a lower assumption for the turbulence as

WEA Class	I	II	III	S
V_{ref} [m/s]	50	42,5	37,5	
V_{ave} [m/s]	10	8,5	7,5	
A I_{15} [–]		0,18		site specific
a		2		
B I_{15} [–]		0,16		
a		3		
C I_{15} [–]		0,145		
a		3		

Fig. 14.1. WEA-classes according to the GL-Offshore guideline [4]

can be seen in Fig. 14.1. For the standard classes a Rayleigh distribution of the wind speed is assumed, for the site specific class a Weibull distribution is used, based on individual measurements. For the FINO1 site Weibull parameters $A = 11.1\,\mathrm{m\,s^{-1}}$ and $k = 2.16$ have been found by an analysis of the data up to January 2005.

14.3 Normal Wind Profile

Preliminary results shall be presented for two parameters of the normal wind conditions. The first one is the normal wind profile (NWP) that is defined quite similar in the above-mentioned guidelines. The mean wind speed V at height z is consistently given as a potential law:

$$V(z) = V_{\mathrm{hub}}(z/z_{\mathrm{hub}})^a, \tag{14.1}$$

where V_{hub} is the wind speed at hub height z_{hub}. Figure 14.2 shows the "onshore" wind profile ($a = 0.2$) according to IEC-61400-1 [6] in comparison with the "offshore" profiles defined by [4,5] ($a = 0.14$). In addition, mean wind profiles as measured at the FINO1 platform during 2004 have been plotted. As could be expected, the "onshore" NWP exaggerates the measured wind profile, while the assumption in the GL and IEC-61400-3 seems to reproduce the right trend, as for the measured mean wind profile it looks like a conservative assumption. The strong dependence of the profile on the atmospheric stability however is neglected. When T_{water} is less than T_{sea}, the measured profile exaggerates the profile of [4,5] drastically. It must be mentioned that these observations are preliminary and must be assured by a more detailed analysis during the OWID project.

14.4 Normal Turbulence Model

Turbulence intensity is one important parameter for the definition of the external load assumptions; normal turbulence is described within the normal wind turbulence model (NTM). It defines a monotone decline of the turbulence

14 Report on the Research Project OWID   83

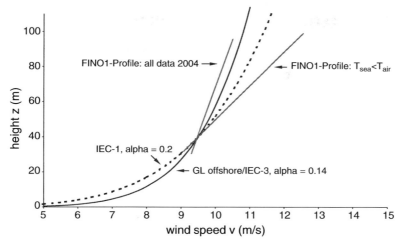

Fig. 14.2. Measured wind profiles at the FINO1 platform in comparison with the IEC-61400-3 and the GL-Offshore guideline

intensity with increasing wind speed. In the GL-guideline the standard deviation of the longitudinal wind speed at hub height σ_1 is defined as:

$$\sigma_1 = I_{15}(15\,\mathrm{m\,s}^{-1} + aV_{\mathrm{hub}})/(a+1) \tag{14.2}$$

with V_{hub} the wind speed at hub height and class parameters a and I_{15} as defined in Fig. 14.1. The turbulence intensity $\sigma_1/V_{\mathrm{hub}}$ is shown in Fig. 14.3

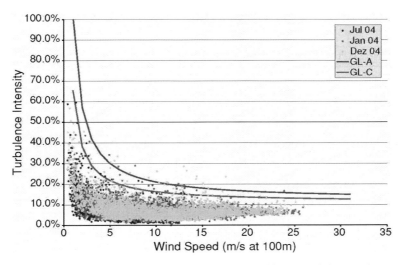

Fig. 14.3. Turbulence intensities according to GL-Offshore guideline and measured at the FINO1 platform within 2004

84 T. Neumann et al.

Table 14.1. Examples for an assessment of extreme wind conditions

No.	Load parameter to be accessed	Data to be assessed from FINO data	Link to standards
1	mean extreme overall load	extrapolated maximum wind speed for a 10 min interval	EWM
2	mean vertical load shear on the rotor	maximum value of vertical wind speed and wind direction shear	EWS
3	Short-term extreme loads	gust parameters depending on height	EOG, EDC
4	time constant for the occurrence of extreme loads	derivation of the maximum wind speed increase during a gust	EOG, EDC

for the turbulence classes A and C. Again measured turbulence intensities at the FINO1 platform have been plotted for comparison. The values have been collected during January, July and December 2004 at the 100 m level.

The intensities in the definition of turbulence class C as well as turbulence class A seem to be an upper limit for the measured values. As a rough picture intensities are scattered around an average value of 5–8% for wind speeds above $8 \, \mathrm{m \, s^{-1}}$, which is only half the value as given by class C. In contrast to the turbulence class model the lowest measured values can be found for wind speeds in between 8 and $10 \, \mathrm{m \, s^{-1}}$ and they are slightly increasing for higher wind speeds, probably because of larger wave heights.

14.5 Extreme Wind Conditions

Extreme wind conditions are defined in order to represent extreme overall loads, extreme load changes and extreme inhomogeneities of the load distribution. They are standardized as:

- Extreme wind speed model (EWM)
- Extreme Operation Gust (EOG)
- Extreme Direction Change (EDC)
- Extreme Coherent Gust (ECG)
- Extreme Coherent Gust incl. Direction Change (ECD)
- Extreme Wind Shear (EWS)

A thorough analysis of the FINO1 data will be carried out in order to verify extreme load conditions for the offshore case. In Table 14.1 examples for the data analysis and the connected extreme wind parameter are given.

14.6 Outlook

The starting point for OWID is the measured "undisturbed" wind conditions at the FINO1 platform. For large offshore wind farms, wind conditions will be severely influenced by the wake fields of the wind turbines. This extra turbulence may dominate the design conditions for certain situations. Therefore in a next step wind park effects will be simulated by using computational fluid dynamics (CFD) methods [7,8]. Both undisturbed and disturbed (park effect) wind conditions are taken as input for modelling a 5 MW wind turbine and study the effect on the layout. Together with an evaluation of life time effects this will be the basis for proposals to improve standard wind conditions in the guidelines.

14.7 Acknowledgement

This work was supported by Federal Ministry for the Environment, Nature Conservation and Nuclear Safety (BMU-Project 0329961a[1]) and ENERCON, GE-Wind Energy, Multibrid, Repower.

References

1. T. Neumann, et al.: DEWI-Magazin Nr. 23, August 2003
2. V. Riedel, et al.: DEWI-Magazin Nr. 26, February 2004
3. Design of Offshore Wind Turbine Structures, Det Norske Veritas, 2004
4. IEC 61400-3, Design Requirements for Offshore Wind Turbines, WD 2005
5. Guideline for the Certification of Offshore Wind Turbines, GL Wind, 2005
6. IEC 61400-1, Design requirements, Ed. 3, 2005
7. Barthelmie, R. (ed.), Risoe National Laboratory, Roskilde, 2002
8. Riedel, V., Diploma Thesis, 2003

---

[1] Design Parameters and Load Assumptions for Offshore WEC in the German Bight on Basis of the FINO-measurements.

15

Simulation of Turbulence, Gusts and Wakes for Load Calculations

Jakob Mann

15.1 Introduction

In order to simulate dynamic loads on a wind turbine rotor it is necessary to simulate the turbulent inflow to the rotor realistically. In this paper we review various methods to do so. In many situations, for example for strong winds over flat terrain, turbulence may be considered for practical load calculation purposes not far from being Gaussian. In this situation it is enough to know the second-order statistics, e.g. spectra and cross-spectra, to be able to simulate. The second-order statistics are either obtained from empirical one-dimensional spectra and coherences as in the Sandia method, or from a semi-empirical three-dimensional spectral tensor as in the Mann model [4,5]. These methods, which both appear in the third edition of the IEC standard 61400-1, will be compared.

The assumptions of the models; stationarity, homogeneity, and gaussianity, can all be questioned, especially when the terrain is not simple or when turbines appear in wind farms. Here a variety of methods are possible and we will present a few practical examples of these. Particular emphasis will be put on "constrained Gaussian simulation" where extreme gusts can be embedded in a turbulent field. This method can also be used simulate entire fields based on measurements in a few points. Practical simulations of wake turbulence based on meandering will also be presented together with preliminary remote sensing measurements [2].

15.2 Simulation over Flat Terrain

The basis of most practical models to simulate the inflow turbulence for aeroelastic codes, such as HAWC and FLEX, is the second order statistics. It is true that real turbulence can not be gaussian. This can, for example, be seen from the famous relation $\langle \delta v_{\parallel}(r)^3 \rangle = -\frac{4}{5}\varepsilon r$ [3], which states that the mean of the cube of difference in the velocity component along the separation

88   J. Mann

vector between two points is $-\frac{4}{5}$ times the dissipation of kinetic energy ε times the separation distance r. It is approximately true for high Reynolds number flows when r is in the inertial subrange, and, assuming an infinitely long inertial subrange, can be derived rigourously. Despite this fact it simplifies the simulation algorithms immensely to assume the velocity field to be a gaussian random field. Furthermore, it is not clear how important the departure from gaussianity is for the dynamic loads, though some attempts to quantify this has been undertaken. For example, for a turbine in rather complex terrain some loads increased by 15% comparing a non-gaussian simulation with a gaussian with the same second order structure [7].

In this paper we from hereon assume gaussianity. We furthermore assume Taylor's hypothesis to be valid

$$\tilde{\boldsymbol{u}}(x, y, z, t) = \tilde{\boldsymbol{u}}(x - Ut, y, z, 0), \tag{15.1}$$

where x is the coordinate in the mean flow direction, U the mean wind speed. The velocity vector is denoted by $\tilde{\boldsymbol{u}}$ and the fluctuations around the mean is $\boldsymbol{u}$. All second order statistics of the fluctuations can be expressed in terms of the covariance tensor

$$R_{ij}(\boldsymbol{r}) = \langle u_i(\boldsymbol{x}) u_j(\boldsymbol{x} + \boldsymbol{r}) \rangle. \tag{15.2}$$

It does not depend on $\boldsymbol{x}$ if homogeneity is assumed. This is usually a good assumption for the horizontal directions x and y, but less optimal for the vertical.

The Fourier transform of the covariance tensor

$$\Phi_{ij}(\boldsymbol{k}) = \frac{1}{(2\pi)^3} \int R_{ij}(\boldsymbol{r}) \exp(-\mathrm{i}\boldsymbol{k} \cdot \boldsymbol{r}) \mathrm{d}\boldsymbol{r}, \tag{15.3}$$

is called the spectral velocity tensor. This has been modelled by Mann [4] for a homogeneous shear layer, and compared well to atmospheric surface-layer data over flat terrain. Based on this model a Fourier simulation algorithm, which essentially is a superposition of sine-waves in 3D space, may be devised [5].

The model by Veers [10], which was improved by Winkelaar [11], is based on one-point spectra and coherences, i.e. the absolute value of the normalized cross-spectra. The phase of the cross-spectra is typically ignored and so is incompressibility. Also, empirical information of all cross-spectra is not available.

A group at Risø has simulated 3D fields by the model and sampled the velocities on a helix emulating the passage of a point on a wind turbine blade through the air [8]. They compared this with measurements made by a pitot tube mounted on a rotating blade. The comparison is shown in the right plot of Fig. 15.1. The left plot in this figure shows the helix sampled spectra from a Veers simulation of only the longitudinal wind component (lower points) and of the two horizontal component. It seems to be important to simulate all three components of the velocity field to obtain realistic spectra.

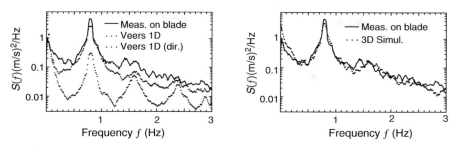

Fig. 15.1. Rotationally sampled turbulence spectra

15.3 Constrained Gaussian Simulation

Special phenomena such as micro-bursts or weird gusts in complex terrain are not taken into account in the simulation described above. Here constrained simulation may be of value. Constrained for wind turbines has been created recently [1] and developed further and formulated differently by Risø.

An example is shown in Fig. 15.2, where a wind increase of 10 m s^{-1} over 20 m (corresponding to a time 20 m/U) in the center of the rotor plane is simulated. The field is random but is required to display the sudden velocity jump. It should be emphasized that the constrained simulation is still Gaussian and that incompressibility is still obeyed [6]. The three-dimensional second-order tensor behind the simulation is described in detail elsewhere [4].

The flexibility of the method is tremendous, but it is not clear how to determine the shape and amplitude of the most critical gusts at a given site.

15.4 Wakes

Another important aspect of dynamic loads is wakes. Turbines in parks often experience large fluctuations in wind speed because an upwind turbine creates a meandering wake. Blades going in and out of this wake can, if the turbines are very close to each other, cause very large loads.

15.4.1 Simulation

We would like to construct simple models of the inflow to turbines in wakes able to produce wind fields suitable for dynamic load calculations.

One such model is shown in Fig. 15.3 where the velocity deficit from eight turbines is advected passively by a large scale turbulent field. This field is simulated according to Mann (1998), but on a quite crude resolution. The wakes expand according to CFD calculation in essentially laminar flow.

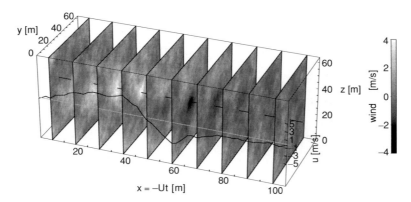

Fig. 15.2. u-turbulence simulation of velocity jump based on the spectral tensor by Mann. Shown in *gray scale* is u in the rotor plane at several several times. The velocity u at the *centreline* where the constraints are made, is also shown on the front of the box. Its maximum is around $5\,\mathrm{m\,s^{-1}}$ and minimum $-5\,\mathrm{m\,s^{-1}}$ as seen from the scale on the front right edge of the box

15.4.2 Scanning Laser Doppler Wake Measurements

To test under which conditions, if any, the wake deficit can be considered as a passively advected quantity we are currently doing an experiment on a small Tellus turbine, with a rotor diameter of 20 m. A wind lidar [9] has been modified to be able to scan the wake downstream of the turbine by mounting it on the rear side of the nacelle. The instrument can remotely measure the component of the wind speed along the laser beam up to a distance of 200 m. By mounting the laser head (see Fig. 15.4) on a movable platform we can make horizontal cuts through the wake deficit.

An example of this is seen in Fig. 15.5. At each downwind distance the lidar scans four times. At the closest distance the wake seems to cover almost the scanning angle. At the other distances the wake appears to the left. As we scan further downstream the wake deficit seems to widen and lessen in amplitude. Scanning at each downstream distance takes of the order of 4 s.

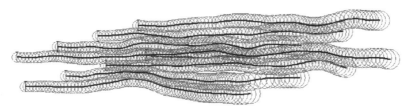

Fig. 15.3. Modelling of wakes advected as passive objects by the large scale environmental turbulence. (M. Nielsen, Risø)

15 Simulation of Turbulence, Gusts and Wakes for Load Calculations

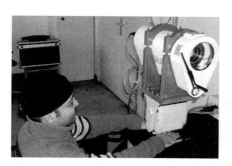

Fig. 15.4. *Left:* The telescope head of the laser Doppler anemometer from QinetiQ [9]. *Right:* Mounting of the laser anemometer behind the nacelle of a Tellus turbine

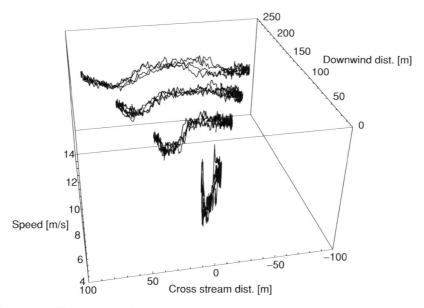

Fig. 15.5. Observation of a wake at four downstream distances: 20, 85, 150, and 215 m. The wake appear to be to the left

The focus is changed in such a way that the same wake deficit is followed downstream from the turbine [2].

More systematic measurements and detailed comparison with the model will soon be done.

92   J. Mann

Acknowledgements

I am grateful to a number of Risø workers for their valuable and helpful contributions. Support from the Danish Energy Authority through grant ENS-33030-0004 is appreciated.

References

1. W. Bierbooms. Investigation of spatial gusts with extreme rise time on the extreme loads of pitch-regulated wind turbines. *Wind Energy*, 8:17–34, 2005
2. F. Bingöl. Adapting a Doppler laser anemometer to wind energy. Master's thesis, Danish Technical University, 2005
3. A. N. Kolmogorov. Dissipation of energy in locally isotropic turbulence. *Dokl. Akad. Nauk SSSR*, 32(1): 1941. English translation in *Proc. R. Soc. Lond. A*, 434:15–17, 1991
4. J. Mann. The spatial structure of neutral atmospheric surface-layer turbulence. *J. Fluid Mech.*, 273:141–168, 1994
5. J. Mann. Wind field simulation. *Prob. Engng. Mech.*, 13(4):269–282, 1998
6. M. Nielsen, G. C. Larsen, J. Mann, S. Ott, Kurt S. Hansen, and Bo Juul Pedersen. Wind simulation for extreme and fatigue loads. Technical Report Risø–R–1437(EN), Risø National Laboratory, 2004
7. M. Nielsen, K. S. Hansen, and B. J. Pedersen. Gyldigheden af antagelsen om Gaussisk turbulens. Technical Report Risø–R–1195(DA), note =In Danish, Risø National Laboratory, 2000
8. J. T. Petersen, A. Kretz, and J. Mann. Importance of transversal turbulence on lifetime predictions for a hawt. In J.L. Tsipouridis, editor, *5th European Wind Energy Association conference and exhibition. EWEC '94*, volume 1, pages 667–673. The European Wind Energy Association, 1995
9. D. A. Smith, M. Harris, A. S. Coffey, T. Mikkelsen, H. E. Jørgensen, J. Mann, and R. Danielian. Wind lidar evaluation at the Danish wind test site in Høvsøre. *Wind Energy*, 9(1–2):87–93, 2006
10. P. S. Veers. Three-dimensional wind simulation. Technical Report SAND88-0152, Sandia National Laboratories, 1988
11. D. Winkelaar. Fast three dimensional wind simulation and the prediction of stochastic blade loads. In *Proceedings from the 10'th ASME Wind Energy Symposium*, Houston, 1991

16

Short Time Prediction of Wind Speeds from Local Measurements

Holger Kantz, Detlef Holstein, Mario Ragwitz, Nikolay K. Vitanov

Summary. We compare different schemes for the short time (few seconds) prediction of local wind speeds in terms of their performance. Special emphasis is laid on the prediction of turbulent gusts, where data driven continuous state Markov chains turn out to be quite successful. A test of their performance by ROC statistics is discussed in detail. Taking into account correlations of several measurement positions in space enhances the predictability. As a striking result, stronger wind gusts possess a better predictability.

16.1 Wind Speed Predictions

Whereas the three dimensional velocity field of the air in the atmosphere can be supposed to be described by the deterministic Navier–Stokes equations, possibly augmented by equations for the temperature and humidity (and hence density of the air) and with suitable boundary conditions, a local measurement yields data which appear to be random. In fact, deterministic behaviour of the local velocities is very unprobable, since (a) the Navier–Stokes equations contain already non-local interactions through the self-generated pressure field, and (b) the local wind speed changes due to the drift of the global wind field across the measurement position.

Data analysis of wind speed recordings yields results which are fully consistent with stochastic data. We use time series recorded in Lammefjord by the Risø research centre [1]. These data are obtained from cup anemometers mounted on measurement masts at heights of 10, 20, and 30 m above ground, taken with 8 Hz sampling rate. The data sets report the absolute values of the wind speeds in the x–y-plane, together with the angle of incidence. In total, we use 10 days of data. Due to the non-stationarity of the data, the autocorrelation function cannot be reliably determined. In any case, correlations of the wind speed decay slowly, whereas the increments, i.e. the differences of succesive measurements, seem to be uncorrelated. Histograms of the differences $v_t - \bar{v}_t$ of the data and the moving 1-minute mean velocities $\bar{v}_t$, conditioned to the value of $\bar{v}_t$, show an almost Gaussian behaviour with variances which

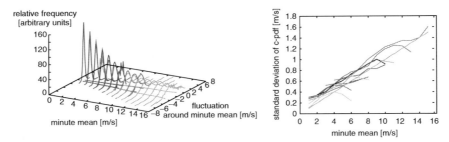

Fig. 16.1. Conditional pdfs $p(v_t - \bar{v}_t|\bar{v}_t)$, where $\bar{v}_t$ is the moving 1-minute mean, of data of three different days and the standard deviations of conditional pdfs as a function of $\bar{v}_t$ for all 10 days

grow linearly with $\bar{v}_t$ Fig. 16.1, so that the noise amplitude of the supposed stochastic process is a function of the state (multiplicative noise). In summary, the very simplest model for wind speeds could read $v_{t+1} = \alpha v_t + \beta v_t \xi_t$, where $0 < \alpha < 1$ and β are parameters and ξ_t is δ-correlated Gaussian noise. Predictions require correlations in data. Different prediction schemes exploit such correlations in different ways. We compare four different predictors which are motivated by the previously discussed properties of the data, written here for predictions k steps ahead:

- persistence: $\hat{v}_{t+k} = v_t$
- extrapolation: $\hat{v}_{t+k} = v_t + (v_t - v_{t-k})$
- linear AR(N)-model: $\hat{v}_{t+1} = \sum_{l=0}^{N-1} a_l v_{t-l}$ k-fold iterated
- order-m Markov chain: $\hat{v}_{t+k} = \int x \, p(x|v_t, v_{t-1}, \ldots, v_{t-m+1}) \, dx$

A contiuous state Markov chain of order m assumes that the data represent a stochastic process which is fully characterized by conditional probability density functions (c-pdf) $p(v_{t+k}|v_t, \ldots, v_{t-m+1})$, i.e. that the knowledge of measurements which are farther in the past than m steps do not influence the probability distribution of the future values [3, 4]. For long range correlated data such a model is an approximation. The conditional probabilities can be estimated from data as follows: Assuming that $p(v_{t+k}|v_t, \ldots, v_{t-m+1})$ is smooth in the conditioning vector, then the "futures" of similar sequences of m measurements are drawn from similar distributions. Hence, if we collect for given time t all past subsequences of length m which are similar to $(v_t, \ldots, v_{t-m+1})$, then their futures form a random sample of the desired conditional distribution. This sample can be used to either estimate $p(v_{t+k}|v_t, \ldots, v_{t-m-1})$ or its moments.

Persistence is expected to be good in cases where increments are uncorrelated and the correlation of the data is strong. The autoregressive model (AR) is expected to be superior if the power spectrum exhibits pronounced peaks. The extrapolation should yield reasonable short time predictions for smooth signals. The Markov chain is able to model a generic non-linear stochastic

16 Short Time Prediction of Wind Speeds from Local Measurements

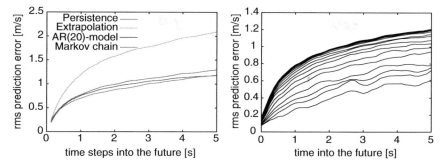

Fig. 16.2. *Left* panel: Prediction errors of different prediction schemes for data from day 191. *Right* panel: Prediction errors of the Markov chain predictor for selected subsamples where the standard deviation of the c-pdf $p(v_{t+1}|v_t,\ldots,v_{t-m+1})$ is smaller than δ, $\delta = 0.25, 0.3, 0.35,\ldots$ from bottom to top

process with short range memory. In the left panel of Fig. 16.2 we show the performance of these predictors on data from a rather turbulent day, as a function of the prediction horizon (all parameters were empirically optimized). The performance is measured in terms of the root mean squared (rms) prediction error, $\bar{e} = \sqrt{\langle(\hat{v}_t - v_t)^2\rangle}$. Since wind speed data are not smooth, predictions by extrapolations are evidently bad. The fact that the three other predictors perform almost equally demonstrates that the simple model $v_{t+1} = \alpha v_t + \xi_t$ with $\alpha \approx 1$ is rather good: Going beyond this hypothesis (which is fully incorporated in the persistence predictor) yields only minor improvements. To account for non-stationarity, the coefficients of the AR(20) model were fitted on moving windows using the past 1,200 s of data. Markov chains of order 10–30 are again slightly superior.

The Markov chain model is flexible enough to model the observed non-constant noise amplitude. Using the variance of the c-pdf as additional information during the forecast process, we can restrict forecasts to those times t when the standard deviation of the actual c-pdf $p(v_{t+k}|v_t,\ldots,v_{t-m+1})$ is smaller than some parameter δ. The rms forecast errors obtained from such subsamples are systematically smaller (Fig. 16.2, right panel). Such an analysis means that we consider our predictions only when we expect them to be good, where, however, the model tells us whether they will be good before the actual prediction error is computed. So the Markov chain yields also a prediction of the state-dependent expected error, which would be a constant for the AR-model.

16.2 Prediction of Wind Gusts

We define the gust of strength g as the rise of the velocity by more than g m s within a time interval of 2 s (16 time steps), where the results are robust

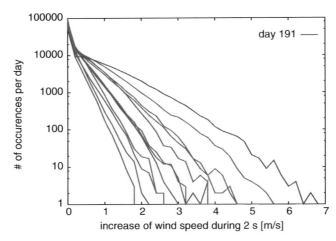

Fig. 16.3. The distribution of wind gusts for different days. Results shown below are obtained for the gusty day 191

against changes of this interval. The histograms of these gust strengthes show approximately an exponential distribution (Fig. 16.3). The previously discussed prediction schemes are not suitable to predict strong gusts, i.e. a coming gust does typically not cause a large positive difference between predicted and actual value, $\hat{v}_{t+k} - v_t$ (the persistence predictor trivially predicts this difference to be 0). However, the Markov chain approach offers a probabilistic way of forecasting: One can extract, for every time instance, a probability of a gust to come. This is done by straightforwardly interpreting the conditional probability densities of the Markov chain or evident variants of these. In order to estimate the probability of a gust to happen during the following 2 s, one selects all data segments at past times $l < t$ which fulfil $\sum_{n=0}^{m-1}(v_{l-n} - v_{t-n})^2 \leq \epsilon^2$, as it was also done to estimate the c-pdf of the Markov chain discussed before. From these "neighbours" we extract the distribution of the maxima of each of the 2s segments following them. The relative number of situations which fulfil $x > v_t + g$, i.e. which fulfil our gust criterion, gives the actual probability of a gust of strength g to come.

This is a probabilistic prediction scheme, and despite a large predicted probability no gust might follow. This illustrates the difficulty of verification: The prediction scheme yields a probability, the actual observation will yield a yes/no result. In order to check the predicted probabilities, we apply the technique of the reliability plot: We create sub-samples of data for which the predicted gust probabilities are in some small interval $p_{\text{gust}}(t) \in [r - \Delta r, r + \Delta r]$. If the predicted probabilities are without any bias, then the relative number of gust events inside such a sample should be in good agreement with r (for sufficiently small Δr). Indeed for r smaller than 1/2 this property is well fulfilled, whereas the deviations at larger probabilities can be explained by statistical fluctuations due to insufficient sub-sample size [6].

The probabilities of a gust to come can be converted into a gust prediction by introducing a threshold p_c: For $p_{\text{gust}}(t) > p_c$ we give a warning, for $p_{\text{gust}}(t) \leq p_c$ we do not. Evidently, the larger p_c, the less frequently a warning will be given. In order to access the performance of this warning scheme, we employ the method of the ROC statistics [7]: We plot the rate of false alarms versus the hit rate. If gust warnings were generated randomly without any correlation to the real future, then the rate of false alarms would be identical to the hit rate, irrespective of p_c and hence irrespective of how frequently we issue a warning. The systematic finding of a higher hit rate than false alarm rate indicates predictive power of the algorithm. Due to the stochastic properties of the wind speeds, error-free predictions are impossible, i.e. for every non-zero hit rate also false alarms occur. The parameter underlying the curves in Fig. 16.4 showing the ROC plot is the threshold value p_c, where $p_c = 1$ corresponds to the origin (no warnings at all), and $p_c = 0$ corresponds to the point $(1, 1)$, maximal hit rate and maximal false alarm rate. The different curves in Fig. 16.4 correspond to results obtained for different gust strengths g. Interestingly, for larger values of g the predictions possess a better hit rate. Meanwhile, we were able to understand the improved predictability for larger gust strengthes in simple model processes [8], which is related to the fact that our "events" are defined as increments with the last observation being the reference.

The Markov chain approach can be easily extended to multi-channel measurements. The conditioning state is then defined by the successive values of all available observables at a number of past time steps. If we do so with measurements which stem from horizontally separated anemometers, then the performance of the predictions is strongly enhanced, if one anemometer is

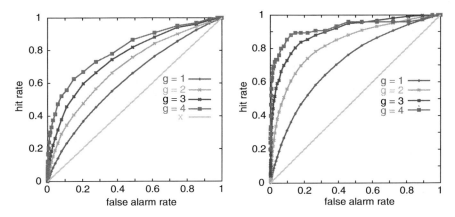

Fig. 16.4. *Left* panel: ROC statistics for univariate time series data. *Right* panel: Improved performance for bivariate input data

98   H. Kantz et al.

up-stream. This can be easily explained by the translation of the wind field across the measurement site. In practice, up-stream measurements would be available only for very selected wind directions. Therefore another finding is very relevant: The gust prediction is also strongly improved if we use bi-variate data recorded at the same mast, namely the wind speeds at 20 and 30 m height above ground, in order to predict the speed at 20 m (Fig. 16.4 right panel). This improvement is absent if we use the same inputs for predictions at the height of the upper, 30 m anemometer. Hence we conclude that there are non-symmetric vertical correlations which supply the information for improved predictions in the lower ranges of the boundary layer. This will be the issue of further investigations.

In summary, predictions of wind speeds based on local measurements on average could not be considerably improved when compared to persistence. However, exploiting additional information contained in the conditional probabilities of a Markov chain model, we could in principle equip every prediction with an expected error bar. The prediction of turbulent gusts in terms of a time dependent risk was more successful, where a conversion of this risk into a warning yields a considerable hit rate for moderate false alarm rates if we focus on high gust strengths and if we use as additional inputs velocity measurements taken at a higher position on the same measurement mast. Similar to many other phenomena, also wind speed data are long range correlated. Since Markov models cannot take into account any long range memory effects, models beyond Markov might yield enhanced predictability.

References

1. Lammefjord data obtained from the Risø National Laboratory in Denmark, http://www.risoe.dk/vea, see also http://www.winddata.com.
2. Box and Jenkins, TIME SERIES ANALYSIS: FORECASTING AND CONTROL, Prentice-Hall, New Jersey, 1994
3. N. G. van Kampen, STOCHASTIC PROCESSES IN PHYSICS AND CHEMISTRY, North-Holland, Amsterdam, 1992
4. F. Paparella, A. Provenzale, L. A. Smith, C. Tarrica, R. Vio. (1997): Local random analogue prediction of nonlinear processes, Phys. Lett. **A 235** 233
5. M. Ragwitz and H. Kantz (2002): Markov models from data by simple nonlinear time series predictors in delay embedding spaces, Phys. Rev. **E 65** 056201
6. H. Kantz, D. Holstein, M. Ragwitz, N.K. Vitanov (2004): Markov chain model for turbulent wind speed data, Physica **A 342** 315
7. K. Fukunaga, INTRODUCTION TO STATISTICAL PATTERN RECOGNITION, Academic, San Diego, 1990
8. S. Hallerberg, E.G. Altmann, D. Holstein, H. Kantz, Precursors of extreme increments, http://arxiv.org/pdf/physics/0604167

17

Wind Extremes and Scales: Multifractal Insights and Empirical Evidence

I. Tchiguirinskaia, D. Schertzer, S. Lovejoy, J.M. Veysseire

Summary. An accurate assessment of wind extremes at various space-time scales (e.g. gusts, tempests, etc.) is of prime importance for a safe and efficient wind energy management. This is particularly true for turbine design and operation, as well as estimates of wind potential and wind farm prospects. We discuss the consequences of the multifractal behaviour of the wind field over a wide range of space-time scales, in particular the fact that its probability tail is apparently a power-law and hence much "fatter" than usually assumed. Extremes are therefore much more frequent than predicted from classical thin tailed probabilities. Storm data at various time scales are used to examine the relevance and limits of the classical theory of extreme values, as well as the prevalence of power-law probability tails.

17.1 Atmospheric Dynamics, Cascades and Statistics

Further to his "poem" [1] presenting a turbulent cascade as the fundamental mechanism of atmospheric dynamics, Richardson [2] showed empirical evidence that a unique scaling regime for atmospheric diffusion holds from centimeters to thousands of kilometers. It took some time to realize that a consequence of a cascade over such a wide range of scale is that the probability tails of velocity and temperature fluctuations are expected to be power-laws [3, 4]. Indeed, this is a rather general outcome of cascade models, independently of their details [5]: the mere repetition of nonlinear interactions all along the cascade yields the probability distributions with the slowest possible fall-offs, i.e. power-laws, often called Pareto laws. The critical and practical importance of the power law exponent q_D of the probability tail, defined by the probability to exceed a given large threshold s, could be understood by the fact that all statistical moments of order $q \geq q_D$ are divergent, i.e. the theoretical moments – denoted by angle brackets $<.>$ – are infinite:

$$\text{for large} \quad s : \Pr(|\Delta v| > s) \approx s^{-q_D} \Leftrightarrow \text{any } q > q_D : \langle |\Delta v|^q \rangle = \infty \qquad (17.1)$$

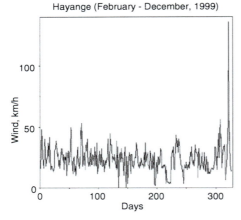
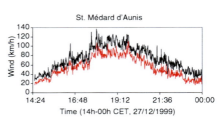

Fig. 17.1a. Daily wind from February to December 1999 at Hayange station

Fig. 17.1b. Mean (red) and maximal (black) wind (both at 1 min resolution) at Saint Médard d'Aunis (4:24–00:00)

and their empirical estimates diverge with the sample size. The probability of having an extreme 10 times larger decreases only by the factor 10^{q_D}, $\kappa=1/q_D$ is often called the "form parameter."

However, by the end of 1950's Richardson's cascade was split [6, 7] into a quasi-2D macro turbulence cascade, a quasi-3D micro turbulence cascade, separated by a meso-scale (energy) gap necessary to avoid contamination of the former by the latter. This gap got some initial empirical support [8], but was more and more questioned [e.g. [9]]. Starting in the 1980's through to the present, this atmospheric model has became untenable thanks to various empirical analyses [10–15] which showed that a new type of strongly anisotropic cascade operates from planetary to dissipation scales. This cascade is neither quasi-2D, nor quasi-3D, but has rather an "elliptical" dimension $D_{el} = 23/9 \approx 2.555$ (in space-time, $29/9 = 3.222$).

Indeed, a scaling anisotropy could be induced by the (vertical) gravity and the resulting buoyancy forces that generate two distinct scaling exponents for horizontal and vertical shears (H_h, respectively, H_v) of the velocity field and the elliptical dimension value is defined by $D_{el} = 2 + H_h/H_v$, the theoretical values $H_h = 1/3$, $H_v = 3/5$ being obtained by a reasoning "à la Kolmogorov" [16] and, respectively, "à la Bolgiano–Obukhov" [17, 18]. The theoretical possibility of having power-law pdf tail also got some empirical support with $q_D \approx 7$ for the velocity field [12, 13].

17.2 Extremes

If wind correlations had only short ranges, then the statistical law of the extremes over a given period would be determined by the classical extreme value theory [19, 20]: the power-law probability tail of the wind series would imply

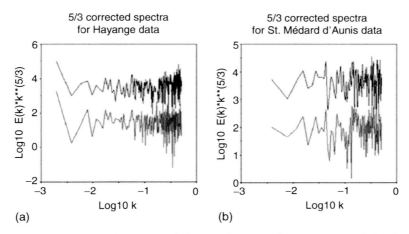

Fig. 17.2. Compensated spectra of the two horizontal components of the data of Fig. 17.1a and of Fig. 17.1b, respectively. The horizontal plateau correspond to the Kolmogorov scaling

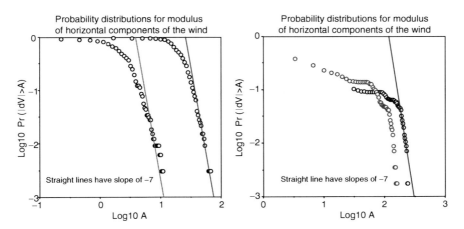

Fig. 17.3. Probability tails of the wind increments of the two horizontal components of the data of Fig. 17.1a and of Fig. 17.1b, respectively

that the wind maxima distribution is a Frechet law [21], often called Generalized Extreme Value distribution of type 2 (GEV2), instead of the more classical Gumbel law (GEV1). GEV1 and GEV2 are quite distinct, since GEV1 has the same power law exponent q_D as the original series, whereas GEV2 has a very thin tail corresponding to a double (negative) exponential. This sharp contrast results from the fact that a Frechet variable corresponds to the exponential of a Gumbel variable, therefore its distribution is often called "Log-Gumbel." We will illustrate this with the help of Météo-France wind data, in particular those collected during the two "inland hurricanes"

102   I. Tchiguirinskaia et al.

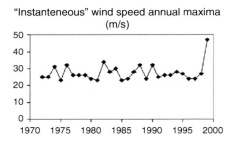

Fig. 17.4. Yearly wind maxima at Montsouris station during the period of 1970–1999

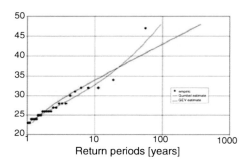

Fig. 17.5. GEV1 (*red straight asymptote*) and GEV2 (*green convex curve*) distributions fitted to the data of Fig. 17.4

that swept across France by the end of the last century (25-26/12/1999 and 27-28/12/1999). They correspond to sharp spikes in daily wind time series at various meteorological stations (Fig. 17.1a). However, intermittent fluctuations are also present at scales of 1 min (Fig 17.1b) and all these fluctuations respect the Kolmogorov scaling (Fig. 17.2 a-b). Furthermore, their probability distributions (Fig. 17.3 a-b) display a rather clear power-law fall-off with an exponent $q_D \approx 7$, in agreement with previous studies.

Looking to larger time scales, let us consider the yearly wind maxima in Paris area during the period of 1970–1999 (Fig. 17.4), as well as the fitted GEV1 and GEV2 distributions (Fig. 17.5) in the so-called Gumbel paper (i.e. wind speed vs. double logarithm of the empirical probability distribution) in which the asymptote of GEV1 is a straight line, whereas GEV2 curve remains convex. It is rather obvious, that both fits are rather equivalent up to the year 1998, whereas the latter can be only captured by the convexity of GEV2.

17.3 Discussion and Conclusion

The hypothesis of short range correlations, which is necessary to derive the classical extreme value theory, is not satisfied by cascade processes. Indeed, the latter introduce power law dependencies. It is therefore indispensable to look for a generalization of extreme value theory in a multifractal framework. This task might be not as difficult as it looks at first glance, since for instance cascade processes may have (statistical) 'mixing' properties [22, 23], which have been used to extend the extreme value theory from uncorrelated time series to those having short range correlations. This may partially explain why GEV2 fits rather well the empirical extremes, although one may expect that its asymptotic power-law exponent might differ from that of the probability tail of the original time series. Larger data bases and numerical cascade simulations are currently being analysed to clarify these issues. However, it may be already timely to use multifractal wind generators in numerical simulations for the design and management of wind turbines.

References

1. Richardson, L.F., *Weather prediction by numerical process.* 1922: Cambridge University Press republished by Dover, 1965
2. Richardson, L.F., *Atmospheric diffusion shown on a distance-neighbour graph.* Proc. Roy. Soc., 1926. **A110**: p. 709–737
3. Schertzer, D. and S. Lovejoy, *The dimension and intermittency of atmospheric dynamics,* in *Turbulent Shear Flow 4,* B. Launder, Editor. 1985, Springer, Berlin Heidelberg New York. p. 7–33
4. Lilly, D.K., *Theoretical predictability of small-scale motions,* in *Turbulence and predictability in geophysical fluid dynamics and climate dynamics,* M. Ghil, R. Benzi, and G. Parisi, Editors. 1985, North Holland, Amsterdam. p. 281–280
5. Schertzer, D. and S. Lovejoy, *Hard and Soft Multifractal processes.* Physica A, 1992. **185**: p. 187–194
6. Monin, A.S., *Weather forecasting as a problem in physics.* 1972, Boston Ma, MIT press
7. Pedlosky, J., *Geophysical fluid Dynamics.* second ed. 1979, Springer, Berlin Heidelberg New York
8. Van der Hoven, I., *Power spectrum of horizontal wind speed in the frequency range from .0007 to 900 cycles per hour.* J. Meteorol., 1957. **14**: p. 160–164
9. Vinnichenko, N.K., *The kinetic energy spectrum in the free atmosphere for 1 second to 5 years.* Tellus, 1969. **22**: p. 158
10. Nastrom, G.D. and K.S. Gage, *A first look at wave number spectra from GASP data.* Tellus, 1983. **35**: p. 383
11. Lilly, D. and E.L. Paterson, *Aircraft measurements of atmospheric kinetic energy spectra.* Tellus, 1983. **35A**: p. 379–382
12. Chigirinskaya, Y., et al., *Unified multifractal atmospheric dynamics tested in the tropics, part I: horizontal scaling and self organized criticality.* Nonlinear Processes in Geophysics, 1994. **1**(2/3): p. 105–114

104   I. Tchiguirinskaia et al.

13. Lazarev, A., et al., *Multifractal Analysis of Tropical Turbulence, part II: Vertical Scaling and Generalized Scale Invariance.* Nonlinear Processes in Geophysics, 1994. **1**(2/3): p. 115–123

14. Lovejoy, S., D. Schertzer, and J.D. Stanway, *Direct Evidence of Multifractal Atmospheric Cascades from Planetary Scales down to 1 km.* Phys. Rev. Letter, 2001. **86**(22): p. 5200–5203

15. Lilley, M., et al., *23/9 dimensional anisotropic scaling of passive admixtures using lidar data of aerosols.* Phys Rev. E, 2004. **70**: p. 036307-1-7

16. Kolmogorov, A.N., *Local structure of turbulence in an incompressible liquid for very large Raynolds numbers.* Proc. Acad. Sci. URSS., Geochem. Sect., 1941. **30**: p. 299–303

17. Bolgiano, R., *Turbulent spectra in a stably stratified atmosphere. J. Geophys. Res.*, 1959. **64**: p. 2226

18. Obukhov, A.N., *Effect of Archimedian forces on the structure of the temperature field in a temperature flow.* Sov. Phys. Dokl., 1959. **125**: p. 1246

19. Gumbel, E.J., *Statistics of the Extremes.* 1958, Colombia University Press, New York 371

20. Leadbetter, M.R., G. Lindgren, and H. Rootzen, *Extremes and related properties of random sequences and processes.* Springer series in statistics. 1983, Springer, Berlin Heidelberg New York

21. Frechet, M., *Sur la loi de probabilité de l'ecart maximum.* Ann. Soc. Math. Polon., 1927. **6**: p. 93–116

22. Loynes, R.M., *Extreme value in uniformly mixing stationary stochastic processes.* Ann. Math. Statist., 1965. **36**: p. 993–999

23. Leadbetter, M.R. and H. Rootzen, *Extremal theory for stochastic processes.* Ann. Prob., 1988. **16**(2): p. 431–478

18

Boundary-Layer Influence on Extreme Events in Stratified Flows over Orography

Karine Leroux and Olivier Eiff

Summary We present a laboratory investigation of the boundary-layer effects on mountain wave-breaking under conditions of uniform flow and stratification, in particular the first complete up- and downstream velocity field revealing upstream separation and blocking, wave breaking, downslope windstorms, trapped lee-waves and rotors. Contrary to the slip condition on the obstacle we show that the slip condition downstream only has a slight influence on the flow dynamics. The results also reveal that the boundary layer developed on the obstacle is controlled by the wave-field, independently of the Reynolds number.

18.1 Introduction

Flow of stably density-stratified fluid over orography, such as mountains, generates internal gravity waves. For strong enough stratifications, i.e. Froude numbers $F_H = U_0/NH < 1$ where U_0 is the wind speed, $N = \sqrt{-\frac{g}{\rho_0}\frac{\partial \rho}{\partial z}}$ is the Brunt–Väisälä frequency and H the mountain height, these waves can attain sufficient amplitude to develop vertical isopycnals which leads to wave-breaking at high altitudes. The ensuing (clear-air) turbulence is accompanied by strong downslope windstorms, with speeds up to $50\,\mathrm{m\,s^{-1}}$, as observed in Colorado on 11 January 1972 [1].

Other orographic phenomena that can occur under stratified conditions are trapped lee-waves (TLW) with embedded rotors. These have usually been associated with inversions and non-uniform flow (e.g. [2]) and are not usually linked to wave-breaking in uniform flow and stratification conditions. However, even for uniform conditions, these phenomena were observed by Eiff et Bonneton [3] when wave breaking occurs. Gheusi et al. [4] showed that the physically correct no-slip condition on the obstacle yields TLWs and rotors even for uniform conditions. In addition to showing that the maximum downslope wind speed is reduced (3–$2.5U_0$), the slip-condition helps to induce

TLWs with embedded rotors and prevents a downslope "shooting" flow. The boundary-layer developed on the obstacle thus plays an important and fundamental role on the wave field dynamics, as was shown for other stratified flow configurations (e.g. [2,5,6]).

18.2 Experimental Procedure

The experiments consist in towing a Gaussian-shaped quasi two-dimensional obstacle in an upright configuration on the bottom of a hydraulic tank at constant velocity U_0, yielding a uniform base flow. The tank is filled with linearly stratified salt water, characterized by a uniform frequency N of order $1\,\mathrm{rad\,s^{-1}}$. (See [3] for more details.) All experiments are carried out at $F_H = 0.6$ where wave breaking occurs. The tank is seeded with particles to apply the particle imaging velocimetry (PIV) technique developed by Fincham and Spedding [7]. The CCD camera and the vertical laser sheet, placed on the symmetry plane parallel to the flow, are fixed to the towing carriage. The no-slip condition behind the obstacle is realized with a thin base-plate attached to the trailing edge of the obstacle. Two tanks, whose dimensions $H_t \times W_t \times L_t$ are $0.7 \times 0.8 \times 7$ m^3 and $1 \times 3 \times 22$ m^3, enable $Re = U_0 H/\nu$ of order 10^2 and 10^4 to be reached, respectively.

18.3 Basic Flow Pattern

Figure 18.1 shows the stationary u-component of the wave field over the obstacle obtained by the PIV technique. (Two separated and overlapping PIV flow-fields have been assembled.) We can clearly identify a wedge of stagnant fluid at high altitude corresponding to the wave-breaking region, below which the flow separates and forms a TLW train. Under both humps, negative velocities are observed, indicative of rotors. The separation occurs at half the wavelength $\lambda = 2\pi U_0/N = 3.8H$ which suggests that the flow separation is lee-wave induced. Given the uniform base N and U_0, TLWs are not expected,

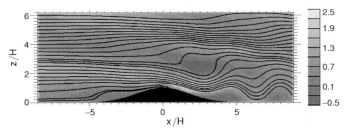

Fig. 18.1. u/U_0 velocity field and streamlines for wave breaking conditions. $Re \sim 10^2$ and no-slip condition is on the obstacle. The flow goes from *left* to *right*

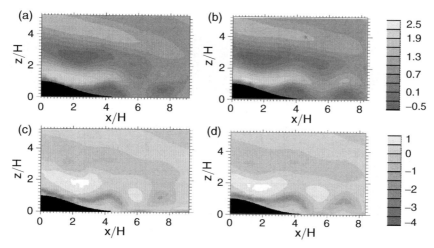

Fig. 18.2. Downstream slip-condition influence. u/U_0 with slip condition (**a**), and no-slip condition (**b**), ω_y/NF_H^{-1} with slip condition (**c**), and no-slip condition (**d**)

but an estimation of two Scorer parameters [8], $l^2 = \frac{gN}{U^2} - \frac{\partial^2 U}{\partial z^2}\frac{1}{U}$, for the perturbed shear flow above the separation, divided into two constant l-layers, is in agreement with the existence of TLWs. On the upstream slope of the obstacle, stagnant fluid with a separated region reaching $z = 0.4H$ can be observed, while further upstream a blocked region of height $\sim 0.2H$, characteristic of low Froude number flows, can be seen.

18.4 Downstream Slip Condition

While the slip condition on the obstacle has been shown to play a major role on the flow dynamics [4], the boundary-condition immediately behind the obstacle is usually free-slip when the obstacles are towed in stationary fluid. Here we introduce a physically correct no-slip condition in the lee of the obstacle itself, at low Re-number. The u/U_0 velocity and ω_y/NF_H^{-1} vorticity fields are presented in Figs. 18.2a and c for the free-slip and in Figs. 18.2b and d for no-slip case, respectively. Contrary to the on-obstacle slip-condition, the downstream-of-the-obstacle slip condition reveals no major influence. It was also observed that the time the waves need to become unstable, break and establish a stationary regime are equivalent in both cases. The lee-side speed increases to $2.5U_0$ at $x = 1.2H$. The reversed flow below the first crest, indicative of a rotor, attains a $0.1U_0$ magnitude, in both cases, even though in Fig. 18.2b it is located just above the slip discontinuity. The vorticity sheets on the first wave crest in Figs. 18.2c and d are significantly higher than the underlying wave vorticity and originate in the boundary layer, clearly indicating a boundary-layer separation that drives the rotor formation. There are

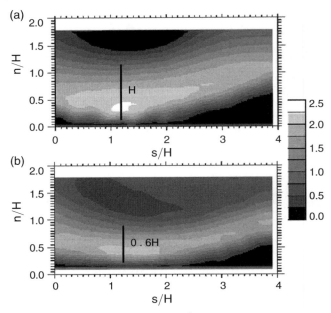

Fig. 18.3. Boundary layer representation. $|\vec{U}|/U_0$ for $Re \sim 10^2$ (**a**), and $Re \sim 10^4$ (**b**)

nevertheless some differences between the two cases. The TLW train of wavelength 0.8λ is shifted downstream by $0.3H$ in the slip case, resulting in a shifted extend of the separated downslope windstorm, in phase with the wave train.

18.5 Boundary Layer and Wave Field Interaction

In Figs. 18.3a and b for $Re \sim 10^2$ and $Re \sim 10^4$, respectively, the norm of the velocity field $|\vec{U}|/U_0$ is plotted on a $(s/H, n/H)$ chart, on the lee side. n is the perpendicular height from surface, and s is the surface-following coordinate starting on the obstacle crest. The thickness of the high velocity region defined as $|\vec{U}| > U_0$ is 40% smaller at high Re-number, but in both cases the maximum velocity is located at a distance of $0.4H$ above the obstacle near $x = \lambda/3$. If the obstacle were a flat plate, the boundary-layer height δ/H would be approximately proportional to $1/H$ for the flow parameters of the two cases. Considering $U_{\max}$ to delimit the boundary-layer height on the obstacles, they do not scale as standard boundary layers, implying, as expected, a strong wave-field influence. The Re-number independence suggests a one-way interaction of the wave field on the boundary-layer height.

18.6 Concluding Remarks

The velocity field up- and downstream of the obstacle has been measured in the laboratory. Contrary to the influence of the on-obstacle slip condition, the after-obstacle slip condition has no major dynamical effect on the flow field. The boundary-layer separation on the obstacle is wave-induced and leads to TLW with embedded rotors. The wave field also appears to control the boundary-layer thickness, independently of the Re-number. Given these results, if the wind-field – including its direction – near the ground and at higher altitudes is to be correctly predicted, the no-slip condition is necessary to be used up to at least the location of the first half internal-wave length.

Acknowledgement

We thank the SPEA team of Météo-France for its strong implication and the PATOM program for its financial support.

References

1. Lilly D K (1978) Am. Meteor. Soc. 35:59–77
2. Doyle J D, Durran D R (2002) J. Atmos. Sci. 59:186–201
3. Eiff O, Bonneton P (2000) Phys. Fluid. 12:1073–1086
4. Gheusi F, Stein J, Eiff O (2000) J. Fluid Mech. 410:67–99
5. Cummins P F (2000) Dyn. Atmos. Ocean. 33:43–72
6. Farmer D, Armi L (1999) Proc. R. Soc. Lond. A. 455:3221–3258
7. Fincham A M, Spedding G R (1997) Exp. Fluids 23:449–462
8. Scorer R S (1949) Quart. J. Roy. Met. Soc. 75:41–56

19

The Statistical Distribution of Turbulence Driven Velocity Extremes in the Atmospheric Boundary Layer – Cartwright/Longuet-Higgins Revised

G.C. Larsen and K.S. Hansen

Summary. We presents an asymptotic expression for the distribution of the largest excursion from the mean level, during an arbitrary recurrence period, based on a "mother" distribution that reflects the exponential-like distribution behaviour of *large* wind speed excursions. This is achieved on the expense of an acceptable distribution fit in the data population regime of small to medium excursions which, however, for an extreme investigation is unimportant. The derived asymptotic distribution is shown to equal a Gumbel EV1 type distribution, and the two distribution parameters are expressed as simple functions of the basic parameters characterizing the stochastic wind speed process in the atmospheric boundary layer.

19.1 Introduction

The statistical distribution of extreme wind speed excursions above a mean level, for a specified recurrence period, is of crucial importance in relation to design of wind sensitive structures. This is particularly true for wind turbine structures.

Assuming the stochastic (wind speed) process to be a Gaussian process, Cartwright and Longuet-Higgens [1] derived an *asymptotic* expression for the distribution of the *largest excursion* from the mean level during an arbitrary recurrence period. From its inception, this celebrated expression has been widely used in wind engineering (as well as in off-shore engineering) – often through definition of the peak factor, which equates the mean of the Cartwright/Longuet-Higgens asymptotic distribution.

However, investigations of full scale wind speed time series, recorded in the atmospheric boundary layer, has revealed that the *Gaussian assumption* is inadequate for wind speed events associated with *large* excursions from the mean [2–4]. Such extreme turbulence excursions seem to occur significantly more frequent than predicted according to the Gaussian assumption, which may under-predict the probability of large turbulence excursions by more than one decade.

19.2 Model

The basic idea behind the model is, in analogy with Cartwright – Longuet-Higgens, to derive an *asymptotic* expression for the distribution of the *largest excursion* from the mean level during an arbitrary recurrence period, however based on a "mother" distribution (different from the Gaussian distribution) that reflects the observed exponential-like behaviour for *large* wind speed excursions. This is achieved on the expense of an acceptable distribution fit in the data population regime of small to medium excursions which, however, for an extreme investigation is considered unimportant. More specifically, we postulate the following *conjecture*: the *tails* of a total population of velocity fluctuations can be approximated by a Gamma distribution with shape parameter $1/2$.

We introduce a (stationary) stochastic wind speed process $U(z,t)$ as

$$U(z,t) = \overline{U}(z) + u(z,t) \ , \tag{19.1}$$

where an upper bar denotes the mean value operator, $u(z,t)$ are turbulence excursions, z denotes the altitude above terrain, and t is the time co-ordinate. The PDF of the *extreme segment* of the excursions, $u_e(z,t)$, is thus expressed as:

$$f_{u_e}(u_e) = \frac{1}{2\sqrt{2\pi C(z)\,\sigma_u}\sqrt{|u_e|}} \exp\left(-\frac{|u_e|}{2C(z)\,\sigma_u}\right) \ , \tag{19.2}$$

where σ_u is the standard deviation of the total data population, and $C(z)$ is a dimensionless, but site- and height-dependant, *positive* constant.

We further introduce the monotone and memory-less transformation, g($*$), defined by

$$v = g(u_e) = \sqrt{\frac{\sigma_u}{C(z)}}\ \text{sign}(u_e)\ \sqrt{|u_e|}\ ; \quad C(z) > 0 \ , \tag{19.3}$$

with the inverse transformation

$$u_e = g^{-1}(v) = \frac{C(z)}{\sigma_u}\ \text{sign}(v)\ v^2\ ; \quad C(z) > 0 \ . \tag{19.4}$$

Formulated in terms of the Gamma PDF, f_{u_e}, the PDF of the transformed variable, f_v, is expressed as

$$f_v(v) = 2\sqrt{\frac{C(z)}{\sigma_u}}\ \sqrt{|g^{-1}(v)|} f_{u_e}\left(g^{-1}(v)\right) = \frac{1}{\sigma_u\sqrt{2\pi}} \exp\left(-\frac{v^2}{2\sigma_u^2}\right) \ , \tag{19.5}$$

which is recognized as a Gaussian distribution.

19 The Statistical Distribution of Turbulence Driven Velocity Extremes   113

As the introduced transformation, $g(*)$, is strictly monotone, every local extreme in the "real" process, $u(z,t)$, is transformed into a local extreme in the "fictitious" transformed process, $v(z,t)$. Thus, the number of local extremes (and their position on the time-axis) is unaltered by the performed transformation.

In the v-domain, the process is Gaussian, and the analysis performed by Cartwright – Longuet-Higgens applies. The derived asymptotic probability density function for the largest maxima, during a time span T, is thus given in terms of excursions normalized with σ_u , η_m, as [1], [5]

$$f_{\max \eta}\left(\eta_\mathrm{m}\right) = |\eta_\mathrm{m}| \, \exp\left(-\mathrm{e}^{-\frac{1}{2}\eta_\mathrm{m}^2 + \ln(Tv)}\right) \mathrm{e}^{-\frac{1}{2}\eta_\mathrm{m}^2 + \ln(Tv)} \, , \tag{19.6}$$

where v is the zero-up-crossing frequency multiplied by 2π.

The final step is to transform this asymptotic result from the v-domain to the u-domain. Denoting ζ as the velocities in the u-domain normalized by σ_u, we finally obtain the requested asymptotic distribution for the largest maxima, ζ_m, as

$$f_{\max \zeta}\left(\zeta_\mathrm{m}\right) = \frac{1}{2C\left(z\right)} \exp\left(-\mathrm{e}^{-\frac{1}{2C(z)}|\zeta_\mathrm{m}| + \ln(Tv)}\right) \mathrm{e}^{-\frac{1}{2C(z)}|\zeta_\mathrm{m}| + \ln(Tv)} \, . \tag{19.7}$$

The asymptotic probability density function expressed in (19.7) is seen to be of the EV1 type [8], which is consistent with the finding from several experimental studies [5–7]. It can be shown that, contrary to the asymptotic PDF described by Cartwright – Longuet-Higgens, the root mean square associated with (19.7) is independent of the time span considered, and the present asymptotic extreme value distribution thus does not narrow with increasing time span.

What remains now is to determine the transformation constant, $C(z)$, such that the best possible agreement between the upper tails of measured PDF's and the proposed asymptotic fit is obtained. This calibration has been performed based extensive full scale measuring campaigns, representing three different terrain types – offshore/coastal, flat homogeneous terrain and hilly scrub terrain.

A common feature of the C-values resulting from the performed data fit, is a moderate, but significant, dependence with height, which may be interpret as the effect of the gradually increasing size of the biggest turbulence eddies contributing to extreme events, as the distance to the blocking ground increases. The following approximation result for the transformation constant has been obtained

$$C\left(z\right) = az + b \, , \tag{19.8}$$

where the values of a and b for the three investigated terrain categories, are given in Table 19.1.

Table 19.1. Estimated parameters defining the transformation constant for various terrains

Terrain type	a	b
Offshore/coastal	0.0014	0.2972
Flat, homogenous	0.0004	0.2820
Hilly, scrub	0.0009	0.3566

References

1. D.E. Cartwright and M.S. Longuet-Higgins (1956). The statistical distribution of the maxima of a random function. Proc. Royal Soc. London Ser. A **237**, 212–232
2. M. Nielsen, G.C. Larsen, J. Mann, S. Ott, K.S. Hansen and B.J. Pedersen (2003). Wind Simulation for Extreme and Fatigue Loads. Risø-R-1437(EN)
3. F. Boettcher, C. Renner, H.-P. Waldl, and J. Peinke (2003). On the statistics of Wind Gusts. Bound. Layer Meteor, **108**, 163–173
4. H.A. Panofsky and J.A. Dutton (1984). Atmospheric Turbulence - Models and Methods for Engineering Applications. Wiley, New York
5. G.C. Larsen and K.S. Hansen (2004). Statistical model of extreme shear. Special topic conference: The science of making torque from wind, Delft (NL), 19–21 April, Delft University of Technology, pp. 433–444
6. G.C. Larsen, K.S. Hansen and B.J. Pedersen (2002). Constrained simulation of critical wind speed gusts by means of wavelets. 2002 Global Windpower Conference and Exhibition, France
7. G.C. Larsen and K.S. Hansen (2001). Statistics of Off Shore Wind Speed Gusts, EWEC'01, Copenhagen, Denmark, 2–6 July
8. Gumbel, E.J. (1966). Statistics of Extremes. Columbia University Express

20

Superposition Model for Atmospheric Turbulence

S. Barth, F. Böttcher and J. Peinke

Summary. We introduce a model that interprets atmospheric increment statistics as a large scale mixture of different subsets of turbulence with statistics known from laboratory experiments. When mixing is weak the same statistics as for homogenous turbulence is recovered while for strong mixing robust intermittency is obtained.

20.1 Introduction

In this paper we focus on the scale dependent statistics of increments $u_\tau = U(t + \tau) - U(t)$ of atmospheric velocities $U(t)$, measured at different on- and offshore locations and compare them to statistics of homogenous, isotropic and stationary turbulence as realized in laboratory experiments. For homogenous, isotropic and stationary turbulence the statistical moments of velocity increments, the so-called structure functions have been intensively studied cf. [1]. Their functional dependence on the scale τ is described by a variety of multifractal models. Besides the analysis of structure functions, probability density functions (PDFs) of the increments are often considered.

The atmospheric PDFs we examine here differ from those of common turbulent laboratory flows where – with decreasing scale τ – a change of shape of the PDFs is observed (e.g. [2]). For large scales the "laboratory" distributions are Gaussian while for small scales they are found to be intermittent. The atmospheric PDFs however change their shape only for the smallest scales and then stay intermittent and non-Gaussian for a broad range of scales, because the outer range is usually unknown or nor accessible. Although the decay of the tails indicates that distributions should approach Gaussian ones (as for isotropic turbulence) they show a rather robust exponential-like decay. The challenge is to describe and to explain the measured fat-tailed distributions and the corresponding non-convergence to Gaussian statistics. Large increment values in the tails directly correspond to an increased probability (risk) to observe large and very large events (gusts).

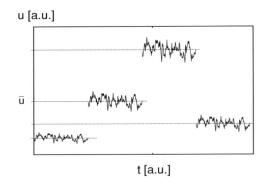

Fig. 20.1. Schematic illustration of different mean velocity intervals. Within these intervals statistics should be the same as for isotropic, homogenous and stationary turbulence. The magnitude of variations (standard deviation) grows with mean velocity

20.2 Superposition Model

We found that the observed intermittent form of PDFs for all examined scales is the result of mixing statistics belonging to different flow situations. These are characterized by different mean velocities as schematically illustrated in Fig. 20.1.

When the analysis by means of increment statistics is conditioned on periods with constant mean velocities results are very similar to those of isotropic turbulence. Note that the definition of the mean velocities requires a separation of scales, which would not exist for a pure fractal process. But the change in statistics we take as hint that there is a separation of scales. This is a controversial issue, which will be tackled further in future.

The change of shape of the PDFs can be described by the *Castaing distribution* that interprets intermittent PDFs $p(u_\tau)$ as a superposition of Gaussian ones $p(u_\tau|\sigma)$ with standard deviation σ. The standard deviation itself is distributed according to a log-normal distribution. The *Castaing distribution* thus reads

$$p(u_\tau|\bar{u}) = \int_0^\infty d\sigma \, \frac{1}{\sigma\sqrt{2\pi}} \exp\left[-\frac{u_\tau^2}{2\sigma^2}\right] \cdot \frac{1}{\sigma\lambda\sqrt{2\pi}} \exp\left[-\frac{\ln^2(\sigma/\sigma_0)}{2\lambda^2}\right]. \quad (20.1)$$

Two parameters enter this formula, namely σ_0 and λ^2. The first is the median of the log-normal distribution, the second its variance. The latter determines the form (shape) of the resulting distribution $p(u_\tau)$ and is therefore called *form parameter*.

We propose a model that describes the robust intermittent atmospheric PDFs as a superposition of isotropic turbulent subsets that are denoted with $p(u_\tau|\bar{u})$ and given by (20.1). Knowing the distribution of the mean velocity $h(\bar{u})$ the PDFs become:

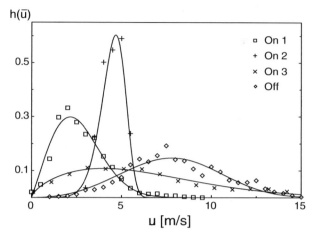

Fig. 20.2. Symbols represent measured mean velocity distributions (averaged over 10 min) of four different atmospheric data sets. *Solid lines* are fits according to (20.3)

$$p(u_\tau) = \int_0^\infty d\bar{u}\, h(\bar{u}) \cdot p(u_\tau|\bar{u}). \tag{20.2}$$

We assume $h(\bar{u})$ to be a Weibull distribution

$$h(\bar{u}) = \frac{k}{A}\left(\frac{\bar{u}}{A}\right)^{k-1} \exp\left[-\left(\frac{\bar{u}}{A}\right)^k\right] \tag{20.3}$$

which is well established in meteorology [3]. In Fig. 20.2 it is shown that a Weibull distribution is a good representation of $h(\bar{u})$.

Inserting (20.3) and (20.1) into (20.2) the following expression for atmospheric PDFs is obtained

$$p(u_\tau) = \frac{k}{2\pi A^k} \int_0^\infty d\bar{u} \int_0^\infty d\sigma\, \bar{u}^{k-1} \exp\left[-\left(\frac{\bar{u}}{A}\right)^k\right]$$
$$\times \frac{1}{\lambda\sigma^2} \exp\left[-\frac{u_\tau^2}{2\sigma^2}\right] \exp\left[-\frac{\ln^2(\sigma/\sigma_0)}{2\lambda^2}\right]. \tag{20.4}$$

Parameters A and k play a similar role as σ_0 and λ^2 in the *Castaing distribution*. With this approach intermittent atmospheric PDFs can be approximated for any location as long as the mean velocity distribution is known.

In Fig. 20.3 probability density functions of velocity increments are shown for different scales τ (increasing from top to bottom). The left figure corresponds to a measurement with an ultrasonic anemometer in a non-stationary atmospheric flow. The right figure corresponds to a measurement with a

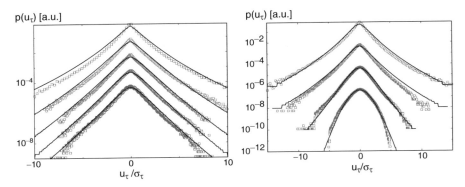

Fig. 20.3. Atmospheric PDFs: Symbols represent the normalized PDFs of the atmospheric data sets. *Straight lines* correspond to a fit of distributions according to (20.4). All graphs are vertically shifted against each other for clarity of presentation. **Left:** sonic anemometer: From top to bottom τ takes the values: 0.5 s, 2.5 s, 25 s, 250 s and 4,000 s. **Right:** hotwire anemometer: From top to bottom τ takes the values: 2 ms, 20 ms, 200 ms and 2,000 ms

hot-wire anemometer in an atmospheric flow during a period of stationary mean wind speed. In both cases the scale dependent statistics can be described with 20.4.

20.3 Conclusions and Outlook

Atmospheric velocity increments and their occurrence statistics are related to loads on wind turbines. For constructing wind turbines a model that reproduces the right increment distributions for every location should therefore be used. Our approach is a good candidate to achieve such a robust model.

References

1. U. Frisch: Turbulence. The Legacy of A.N. Kolmogorov, Cambridge University Press, Cambridge, 1995
2. B. Castaing, Y. Gagne, E.J. Hopfinger: *Physica D* **46**, 177, 1990
3. T. Burton, D. Sharpe, N. Jenkins and E. Bossanyi: Wind Energy Handbook, Wiley, New York, 2001

21

Extreme Events Under Low-Frequency Wind Speed Variability and Wind Energy Generation

Alin A. Cârsteanu and Jorge J. Castro

Summary. Low-frequency wind speed variability represents an important challenge to statistical estimation for atmospheric turbulence and its impact on wind energy generation, given that it does not allow for the usual stationarity assumption in time series analysis. This work presents a framework for the parameterization of the cascade representation of turbulent processes, where stationarity of the cascade generator is assessed from the breakdown coefficients of time series.

21.1 Introduction

The effect of low-frequency (or climatic-scale) variability in wind velocities amounts to nonstationarity in the time series at scales comparable to the series length. When this is the case, the probabilities of future events cannot be inferred statistically from the past. In particular, the estimated probabilities of extreme events, which are by definition scarce and therefore difficult to estimate from statistics in the absence of phenomenologically based models even under stationary conditions, are the most sensitive to nonstationarities.

On the other hand, empirical observations have recently triggered an alarm concerning more frequent atmospheric extreme events, associated with climatic-scale variations (in lay terms, climate change), be they of human or natural origin. This raises the problem of being able to quantify a variation in the parameters of the underlying probability distribution functions of those extreme events, without using directly the statistics of extreme events, which are unable to offer reasonable confidence levels due to their scarcity. We propose here a scaling stationarity criterion, based on the multifractal nature of energy cascading in the terrestrial atmosphere.

Atmospheric extreme events are typically associated with the occurrence of extreme velocity departures of turbulent fluctuations over contiguous space-time domains, a phenomenon called "coherence." These coherent structures are clustering effects typical of the multifractal energy cascade which governs turbulent velocity scaling. The scaling law of velocity fluctuations,

120   A.A. Cârsteanu and J.J. Castro

which, although not yet derived from prime principles, has sample phenom-enological basis and experimental verification (see e.g. [4]), is the so-called "structure function" $\zeta(p)$: $|\Delta v(\Delta t)|^p := \langle |v(t + \Delta t) - v(t)|^p \rangle_t \propto \Delta t^{\zeta(p)}$. To transpose this relationship from temporal to spatial scaling, Taylor's "frozen-field" hypothesis [9], stating that the intrinsic temporal partial derivative of velocity is negligible with respect to the inertial derivative, is currently widely used. While undoubtedly verified in high-speed wind-tunnel experiments, the applicability of the hypothesis in atmospheric conditions is questionable [3,6]. However, in the present work, we shall limit ourselves to considering temporal scaling issues.

21.2 Mathematical Background

In the context of the large-deviations property of multifractal fields, the well-known (see e.g. [1, 5, 8]) asymptotic behavior of probability distribution functions with scale, in the small-scale, large-intensity limit, was recast in [2] as: $\lim_{\lambda \to \infty} g(\lambda) \lambda^{c(\gamma)} P(\Delta v(\Delta t/\lambda) v_0 \lambda^\gamma) / P(\Delta v(\Delta t) > v_0 \lambda^\gamma) = 1$, where λ is a scale factor, v_0 is some constant velocity, γ is an order of singularity, $c(\gamma)$ is the corresponding codimension function of the multifractal field, related to the above-mentioned structure function through a Legendre transform, and $g(\lambda)$ is a slowly varying function of λ, in the sense that $\lim_{\lambda \to \infty} g(a\lambda)/g(\lambda) = 1$, $\forall a \in \mathbf{R}^+$.

Since probabilities are understood here in the temporal sense, rather than across the ensemble (let us notice that in the general case, multifractal fields are trivially nonergodic), at least stationarity should be achieved, in a sense that would allow us to parameterize probability distributions of future events from past events. However, no such stationarity is present in general in mul-tifractal fields. Moreover, due to the divergence of moments of $\Delta v(\Delta t)$, there exists a whole domain of orders of singularity where the codimension function $c(\gamma)$ is a linear function of γ. This domain corresponds precisely to the highest values of orders of singularity up to $\gamma_{\max}$. This is to say that the worst-case scenario, in the sense of the lowest return period corresponding to a given extreme event, is reached over this domain, due to the fact that the highest orders of singularity dominate the field at small scales. This same property implies that if a given sample does not capture the linear interval of $c(\gamma)$, then the estimated value of $\gamma_{\max}$ will not result in the true highest value of the exponent, but rather in a sample-dependent value, which is where nonsta-tionarity of the multifractal field is manifest in the sense of extreme values. Therefore, we propose to define (and statistically test) a stationarity in the scaling characteristics of the field, in such a way that the stationarity of the field generator can be assessed, and with it, the stationarity of the probabilities of extreme events.

Breakdown coefficients have been defined ([7]) for scalar fields (such as e.g. energy E), as: $B_{i,j}(\lambda) := E(\lambda \Delta t_i)/E(\Delta t_j)$, where the indexing i, j occurs along the support. For scale-invariant fields, the statistics of B depend only on the scale ratio λ and not on the absolute scale Δt. Although breakdown coefficients do not assume any particular model for the analyzed field, their results may be used to characterize and evaluate parameters of scaling models, such as multiplicative cascades. It should be noted that breakdown coefficients represent the measurable values of the weights in a multiplicative cascade model, which are not identical to the generating cascade weights, except in the restrictive case of a microcanonical cascading process. The fact that breakdown coefficients only depend on the weights' generator in the case of a multiplicative cascade [3], allows for their use not only for scale-invariance testing, but also for generator stationarity probing. This is to say that, while multifractal fields are, as such, typically nonstationary, we can nevertheless unveil the possible stationarity of the field generator, and make use of it for prediction purposes.

21.3 Results and Conclusions

The data used for this study comprise wind velocities measured at a deforested site in the state of Rondônia, Brazil (10°45′S, 62°22′W) during January and February 1999. The land is dominated by short grasses, about 25 cm tall; isolated indigenous trees are scattered throughout the landscape. The measurements were obtained as part of the NASA TRMM-LBA (Tropical Rainfall Measuring Mission – Large scale Biosphere-Atmosphere) project in the Amazonia. To measure the three components of the wind speed, a sonic anemometer (Solent A1002R, Gill Instruments, Lymington, UK) was deployed on a tower at 6 m above the surface. The sonic anemometer recorded wind speed at a frequency of 10 Hz. These fast-response data were obtained via data acquisition and electronic signal condition systems (model SCXI 2400, National Instruments, Austin, TX) which were interfaced with a computer.

The breakdown coefficient analysis has been performed on daily data, in order to distinguish whether the energy cascading process can be regarded as similar between day and night. While the absolute values of velocities certainly look very different, the breakdown coefficient distributions of energy dissipation (Fig. 21.1) are very similar, and a Kolmogorov-Smirnov test indicates that the hypothesis of the two empirical distributions having the same underlying probability distribution function can be accepted at high significance levels (e.g. 20%).

We conclude therefore that the multifractal cascade generator of energy dissipation in the atmosphere can be regarded as stationary (which implies stationarity of the fluctuations field up to a multiplicative intensity modulation), in this case between day and night, with implications on the mechanism underlying Taylor's hypothesis that are being discussed elsewhere [3]. The same method of breakdown coefficient analysis can be used to determine the stationarity of atmospheric turbulent energy cascade generators at larger temporal scales, thereby drawing conclusions on the interplay of climatic indexes, possible climate change, and others.

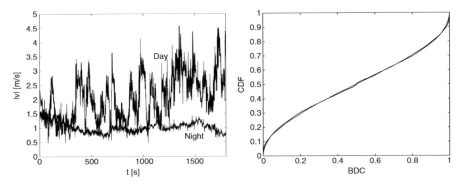

Fig. 21.1. Absolute values of wind speed at night (starting at 1:00 a.m., marked "Night") and during daytime (starting at 12:00 noon, marked "Day") on January 28, 1999 (*left*), and the empirical cumulative distribution functions (CDF) of breakdown coefficients (BDC) of energy dissipation during daytime and at night (*right*)

21.4 Acknowledgments

The authors gratefully acknowledge SEMARNAT-CONACyT grants C01-0615 and C01-0306, as well as NASA's Tropical Rainfall Measurement Mission. A kind recognition goes to the organizers of the EuroMech Wind Energy Colloquium at the Carl von Ossietzky University of Oldenburg, Germany.

References

1. Burlando P., Rosso R. (1996) Scaling and multiscaling models of depth-duration-frequency curves for storm precipitation. J Hydrol 187:45–64
2. Castro J.J., Cârsteanu A.A., Flores C.G. (2004) Intensity-duration-area-frequency functions for precipitation in a multifractal framework. Physica A 338:206–210
3. Cârsteanu A.A., Castro J.J., Fuentes J.D. (2004) Atmospheric turbulence structure and precipitation occurrence in a tropical climate. Eos Trans AGU 85(17):NG21A-03
4. Frisch U. (1995) Turbulence. Cambridge University Press, Cambridge
5. Hubert P., Tessier Y., Lovejoy S., Schertzer D., Schmitt F., Ladoy P., Carbonnel J.P., Violette S., Desurosne I. (1993) Multifractals and extreme rainfall events. Geophys Res Lett 20(10):931–934
6. Lindborg E. (1999) Can the atmospheric kinetic energy spectrum be explained by two-dimensional turbulence? J Fluid Mech 388:259–288
7. Novikov E.A. (1994): Infinitely divisible distributions in turbulence. Phys Rev E 50(5):R3303–R3305
8. Schertzer D., Lovejoy S. (1987) Physical modeling and analysis of rain and clouds by anisotropic scaling multiplicative processes. J Geophys Res 92D(8):9693–9714
9. Taylor G.I. (1938) The spectrum of turbulence. Proc R Soc Lond A 164:476–490

22

Stochastic Small-Scale Modelling of Turbulent Wind Time Series

Jochen Cleve and Martin Greiner

Summary. Today's wind field simulation tools are based on Gaussian statistics and if they resolve the smallest scales they are not concerned about a consistent description of velocity and dissipation. We present a data-driven stochastic model that provides such a consistent description. The model is a multifractal extension of fractional Brownian motion, it also describes the non-Gaussian statistics of turbulent wind fields. In order to further integrate skewness and stationarity an additional small correlation between the multifractal part and the fractional Brownian motion is added.

22.1 Introduction

A sound understanding and modelling of small-scale turbulence in wind fields is – besides being an interesting and challenging problem in itself – important in many real world applications. For example, the design of modern wind energy converters requires a detailed modelling of the flow around the rotor blades. Moreover, with a better understanding of turbulent flow properties an ultra-short prognosis might become feasible which would be a beneficial contribution for an intelligent condition monitoring and controlling.

There are plenty of stochastic models that can reproduce selected statistical properties of a small-scale turbulent velocity field [3]. All of these do not include proper features of the dissipation ε. Likewise good models for the dissipation exist [6], but it is not straightforward to translate dissipation into a velocity field. We will present a data-driven phenomenological model that provides a consistent description of the velocity as well as the dissipation field.

22.2 Consistent Modelling of Velocity and Dissipation

A fully-developed turbulent velocity field is similar in many aspects to fractional Brownian motion (fBm), but obviously turbulence is more complex. On the other hand phenomenological modelling of the dissipation ε has been

124   J. Cleve and M. Greiner

accomplished very successfully by means of random multiplicative cascade processes (RMCP) [5–7]. The basic idea of our model is to extend fBM with multifractal properties of RMCPs.

fBM $B_H(t)$ is an extension of ordinary Brownian motion where the variance of increments $\langle (B_H(t) - B_H(t'))^2 \rangle \sim (t - t')^{2H}$ have a Hurst exponent $0 \leq H \leq 1$ other than $H = \frac{1}{2}$; see e.g. [2] for details. Regarding RMCPs we pick the most successful variant [6], where the dissipation $\varepsilon(x,t) = \exp\left\{ \int_{t-T}^{t} dt' \int_{x-g(t-t')}^{x+g(t-t')} dx'\, \gamma(x',t') \right\}$ is defined as a causal integral over an independently, identically distributed Lévy-stable white-noise field $\gamma(x,t) \sim S_\alpha((dxdt)^{\alpha^{-1}-1}\sigma, -1, \mu)$ with index $0 \leq \alpha \leq 2$. The function $g(t) = \frac{1}{2}\frac{LT\eta}{L(T-t)}$ is defined such that a desired universal multifractal model correlation structure of the dissipation is established. L and T are a correlation length and time, respectively.

Because energy dissipation is proportional to the square of velocity derivatives our process is defined as the product

$$\frac{\partial u}{\partial t} = M_\varepsilon \Delta B_H \qquad (22.1)$$

of a multifractal field $M_\varepsilon \propto \sqrt{\varepsilon}$ and an incremental fBm ΔB_H with Hurst exponent $H = 1/3$. The process (22.1) is a direct generalisation of the multifractal random walk proposed in [1]. Note that the ansatz (22.1) multiplies two fields defined at the dissipation scale. Although beyond the scope of this contribution, it would be very interesting to compare this approach with an implementation where velocity increments associated with an energy flux evolve from the integral scale down to the dissipation scale, which appears closer to the intuitive physical picture of a turbulent energy cascade, see e.g. [3].

With this ansatz it is guaranteed that the correlation structure of the dissipation $\varepsilon \propto (\partial u/\partial t)^2 \propto (M_\varepsilon \Delta B_H)^2$ is determined solely by the RMCP. Because M_ε and ΔB_H are uncorrelated, the two-point correlation function $\langle \varepsilon(x_1)\varepsilon(x_2) \rangle \propto \langle (M_\varepsilon(x_1))^2 (M_\varepsilon(x_2))^2 \rangle \langle (\Delta B_H(x_1))^2 (\Delta B_H(x_2))^2 \rangle$ factorises. The second factor does not depend on the two-point distance $x_2 - x_1$.

Similarly, the properties of the velocity field are given predominantly by the properties of fBm. The standard way to characterise the statistics of the velocity field are structure functions $S_n(r) = \langle (u(x+r) - u(x))^n \rangle \propto r^{\zeta_n}$. Plain fBm would result in the linear spectrum of K41 theory. The M_ε-extension leads to the non-linear spectrum of the scaling exponents.

22.3 Refined Modelling: Stationarity and Skewness

The simple combination (22.1) of the two processes would fall short of modelling turbulence for at least two reasons. fBm is a non-stationary and symmetric process whereas turbulence is stationary and skewed. A generalisation of multifractal fractional Brownian motion is needed. Guidance comes

22 Stochastic Small-Scale Modelling of Turbulent Wind Time Series

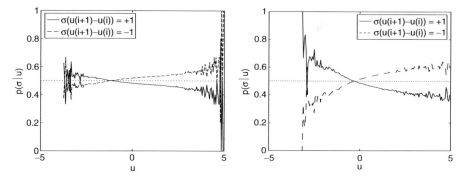

Fig. 22.1. PDF of σ conditioned on the current velocity value for a real turbulent flow (*left*) and the process (22.2) (*right*)

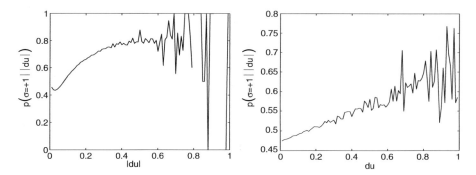

Fig. 22.2. Probability that an increment of size du is going to the positive direction for a real turbulent flow (*left*) and the process (22.2) (*right*)

from an investigation of turbulent data recorded in a wind tunnel [4] and leads to focus on sign statistics.

Skewness unravels in the non-zero odd-order structure functions and the asymmetry of the probability density functions (PDFs) $p(u)$ of the velocity field and $p(du)$ of velocity increments. This asymmetry can be depicted nicely in conditional sign-probabilities. Figure 22.1a shows the probability $p(\sigma|u)$ of the sign σ of the next increment conditioned on the present value of the velocity. Note that the turning point u_{tp}, where it is equally likely to make a step to the positive or to the negative direction is not the mean velocity, which by normalisation has been set to $\langle u \rangle = 0$. The sign probability $p(\sigma|du)$ conditioned on the increment size is shown in Fig. 22.2a. This PDF demonstrates that a big increment is more likely to occur in the positive direction than in the negative direction. These two asymmetries play together such that the mean $\langle du \rangle = 0$ is zero.

126   J. Cleve and M. Greiner

To incorporate the observed asymmetry and stationarity into our model we modify the fBM by introducing small correlations between the fBm process and the multifractal process

$$\frac{\partial u}{\partial t} = M_\varepsilon \Delta \hat{B}_H(u, \mathrm{d}u).$$ (22.2)

The modifications are implemented simply by changing the sign σ of the fBm with the product of the probabilities $P_{\sigma \to -\sigma}(u)$ and $P_{\sigma \to -\sigma}(\mathrm{d}u)$. As a first approximation the two probabilities are chosen as:

$$P_{\sigma \to -\sigma}(u) = \begin{cases} \min(a(u + u_{\mathrm{tp}}), 1) & u > u_{\mathrm{tp}} \wedge \sigma = 1 \\ \min(|a(u + u_{\mathrm{tp}})|, 1) & u < u_{\mathrm{tp}} \wedge \sigma = -1 \\ 0 & \text{otherwise} \end{cases} ,$$ (22.3)

$$P_{\sigma \to -\sigma}(\mathrm{d}u) = \begin{cases} \min(|b\mathrm{d}u|, 1) & \sigma = -1 \\ 0 & \sigma = +1 \end{cases} .$$ (22.4)

For the simulation the parameters have been set to $u_{\mathrm{tp}} = -1$, $a = 0.05$ and $b = 0.4$. This clones the observed correlations between σ, u and $\mathrm{d}u$. For example, the first probability ensures stationarity by turning a step of the fBm which would lead further away from u_{tp} back towards u_{tp} with a small likelihood proportional to the distance from u_{tp}.

22.4 Statistics of the Artificial Velocity Signal

With the correlations introduced by (22.3) and (22.4) the process (22.2) generates a signal that resembles a fully developed turbulent velocity field in every detail. The sign-statistics $p(\sigma|u)$ and $p(\sigma|\mathrm{d}u)$ of such an artificial velocity signal are shown in Figs. 22.1b and 22.2b. The asymmetry apparent in these correlations is mirrored in skewed PDFs $p(u)$ and $p(\mathrm{d}u)$ and also changes the behaviour of the odd-order structure functions, which now also show scaling with scaling exponents close to the ones obtained from experimental data. We have cross-checked that the modifications of the fBm do not alter the scaling properties of the dissipation field.

With the combination of fBm and multifractal RMCP (and a small correlation between them) it is possible to model a turbulent velocity field which has qualitatively the same statistical properties as a real turbulent flow. Furthermore, the signal yields consistent dissipation statistics. The presented scheme can be used to model close to reality wind fields and is easily extendable to higher dimensions.

References

1. E. Bacry, J. Delour, and J.F. Muzy. Multifractal random walk. *Phys. Rev. E*, 64:026103, 2001
2. J. Feder. *Fractals*. Plenum, New York, 1988

22 Stochastic Small-Scale Modelling of Turbulent Wind Time Series

3. A. Juneja, D.P. Lathrop, K.R. Sreenivasan, and G. Stolovitzky. Synthetic turbulence. *Phys. Rev. E*, 49(6):5179–5194, 1994
4. B.R. Pearson, P.A. Krogstad, and W. van de Water. *Phys. Fluids*, 14:1288, 2002
5. D. Schertzer and S. Lovejoy. *J. Geophys. Res.*, 92:9693, 1987
6. J. Schmiegel, J. Cleve, H. Eggers, B. Pearson, and M. Greiner. Stochastic energy-cascade model for (1+1) dimensional fully developed turbulence. *Phys. Lett. A*, 320:247–253, 2004
7. F.G. Schmitt and D. Marsan. *Eur. Phys. J. B*, 20:3, 2001

23

Quantitative Estimation of Drift and Diffusion Functions from Time Series Data

David Kleinhans and Rudolf Friedrich

Summary. This contribution provides an introduction to the concept of drift and diffusion functions for complex dynamical systems such as wind energy converters. These functions easily can be estimated from measured data. However, one has to be aware about intrinsic errors in the estimation procedure that are discussed in the following.

23.1 Introduction

Researchers in the field of the construction of wind energy converters are confronted with a complex problem: the number of degrees of freedom of the wind turbine is extraordinary high. In addition to the adjustable parameters of the rigid body such as the pitch of the rotor-blades many dynamical modes of different parts of the converters have to be considered for a complete description. Moreover the incoming flow is turbulent and fluctuating in space as well as in time.

Complex behaviour in systems far from equilibrium can quite often be traced back to rather simple laws due to the existence of processes of self-organization. For adequate order parameters the dynamics is determined by stochastic differential equations incorporating deterministic as well as stochastic forces. Knowledge of the deterministic part of the dynamics can lead to a deeper understanding of the properties of the system under consideration while the stochastic forces account for the effects of the fluctuating microscopic degrees of freedom. For certain order parameters these forces have properties that are well-known in the theory of stochastic processes. Imagine for example the power output of a wind turbine. Usually the power output is investigated as a function of the (mean) wind speed. Without a doubt much more parameters of the turbine effect the output power and the dynamics of the system act much faster than the common averaging periods.

Recently it has become evident, that knowledge of the stochastic dynamics has significant advantages with respect to the conventional wind power curves:

130   D. Kleinhans and R. Friedrich

Anahua et al. considered the power output of the turbine as a stochastic process [1]. A standard method allows for direct estimation of the characteristic drift and diffusion functions from measured data [2]. By means of this method Anahua et al. could extract the real time dynamics and in the meantime are able to detect small deviations of the control system of the turbine from the optimal working state.

23.2 Direct Estimation of Drift and Diffusion

Generally one has to distinguish between dynamical and measurement noise: measurement noise is superimposed to the data during the measurement process and has no further influence on the system's dynamics. On the other hand many complex systems on a macroscopic scale show some intrinsic, dynamical noise stemming from the microscopic degrees of freedom. Under certain conditions the time evolution of the state $\boldsymbol{x}$ of such systems can be described by Langevin equations of the type:

$$\dot{\boldsymbol{x}} = \boldsymbol{D^{(1)}}(\boldsymbol{x}) + \sqrt{D^{(2)}(\boldsymbol{x})}\boldsymbol{\Gamma}(t). \tag{23.1}$$

$\boldsymbol{\Gamma}$ represents an independent, delta correlated and normal distributed stochastic force that obeys $\langle \Gamma_i(t)\Gamma_j(t')\rangle = 2\delta_{ij}\delta(t-t')$. We apply Itô's interpretation of stochastic integrals [3]. The corresponding Fokker-Planck equation (FPe) characterizes the evolution of the probability density function (pdf) f with time,

$$\frac{\partial}{\partial t} f(\boldsymbol{x},t) = \left(-\sum_i \frac{\partial}{\partial x_i} D_i^{(1)}(\boldsymbol{x}) + \sum_{ij} \frac{\partial^2}{\partial x_i \partial x_j} D_{ij}^{(2)}(\boldsymbol{x}) \right) f(\boldsymbol{x},t). \tag{23.2}$$

$\boldsymbol{D^{(1)}}(\boldsymbol{x})$ is called the drift vector, $D^{(2)}(\boldsymbol{x})$ the diffusion matrix of the corresponding stochastic system.

From the Kramers-Moyal expansion [3] – the more general origin of the Fokker-Planck equation that covers non-Markovian processes – the following definition is known:

$$D^{(n)}(\boldsymbol{x}) := \frac{1}{n!} \lim_{\tau \to 0} \frac{1}{\tau} \langle [\boldsymbol{x}(t+\tau) - \boldsymbol{x}(t)]^n \,|\boldsymbol{x}(t) = \boldsymbol{x}\rangle. \tag{23.3}$$

It has been shown [2] that this expression applied for $n = 1$ and $n = 2$ can be used for direct estimation of drift and diffusion functions, respectively, from time series data. This procedure successfully has been applied to the power output of wind turbines [1] and various problems in medical and life science. The computational requirements for this method are outstandingly low. However, the required discretization of state space and – in particular – the limiting procedure with respect to the time increment makes high demands on the time series data with regard to the sampling frequency and the amount of data-points. From now on the estimation of an one-dimensional process is discussed. The generalization to higher dimensions follows accordingly.

23.3 Stability of the Limiting Procedure

Any measured time series has a finite sampling rate that limits the available time increments for the limiting procedure (23.3). Hence this expression has to be extrapolated to the value $\tau \equiv 0$.

A formal solution of the FPe for the conditional pdf $p(x, t|x_0, t_0)$ is

$$p(x, t|x_0, t_0) = \exp\left[\hat{L}\,(t - t_0)\right]\delta(x - x_0) \tag{23.4}$$

with $\hat{L}$ being the Fokker-Planck operator. A Taylor expansion of this expression yields

$$p(x, t_0 + \tau|x_0, t_0) = \left(1 + \tau\hat{L} + \frac{\tau^2}{2}\hat{L}^2 + \mathcal{O}(\tau^3)\right)\delta(x - x_0). \tag{23.5}$$

This pdf can be used for analytical calculation of the conditional moments (23.3). Eventually one can assess the deviations of the estimate of drift and diffusion $D_{\mathrm{E}}^{(i)}(x, \tau)$ for finite τ from the intrinsic functions $D^{(i)}(x)$. The first order corrections read:

$$D_{\mathrm{E}}^{(1)}(x, \tau) \approx D^{(1)}(x) + \frac{\tau}{2}\left[D^{(1)}(x)\frac{\partial}{\partial x}D^{(1)}(x) + D^{(2)}(x)\frac{\partial^2}{\partial x^2}D^{(1)}(x)\right]$$

$$D_{\mathrm{E}}^{(2)}(x, \tau) \approx D^{(2)}(x) + \frac{\tau}{2}\left[D^{(1)}(x)D^{(1)}(x) + 2D^{(2)}(x)\frac{\partial}{\partial x}D^{(1)}(x) \right. \tag{23.6a}$$

$$\left. + D^{(1)}(x)\frac{\partial}{\partial x}D^{(2)}(x) + D^{(2)}(x)\frac{\partial^2}{\partial x^2}D^{(2)}(x)\right]. \tag{23.6b}$$

Depending on the shape of drift and diffusion functions significant deviations from the intrinsic functions occur for finite τ. These deviations cannot be grasped with statistical considerations as they originate in the properties of the propagator for finite time.

23.4 Finite Length of Time Series

On the other hand one has to consider the finite number of data points. Discretization of state space confines the number of points even more. Especially in sparsely populated regions that come along with natural boundary conditions the low density leads to huge errors in the estimation of drift and diffusion.

A suitable measure for the error margin of the estimate is the standard deviation, the root mean square displacement of the increments from their mean. If N measurements contribute to the averages, the resulting error $E_N\left[D_{\mathrm{E}}^{(1)}(x_0, \tau)\right]$ in the estimated drift function gets:

132   D. Kleinhans and R. Friedrich

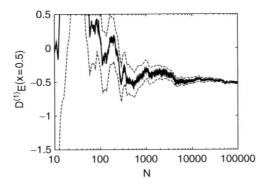

Fig. 23.1. Estimated drift function $D_E^{(1)}(x=0.5)$ of synthetic Ornstein-Uhlenbeck process as function of N. *Dashed region* marks the symmetric error-bars corresponding to (23.7)

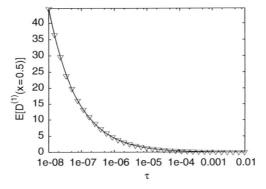

Fig. 23.2. Error of drift estimate as function of time increment τ (*triangles*). A divergent behaviour for $\tau \to 0$ is evident. The *solid line* represents the best fit $f(\tau) = 0.0045/\sqrt{\tau}$

$$E_N\left[D_E^{(1)}(x_0,\tau)\right] = \sqrt{\frac{2}{\tau}\frac{D_E^{(2)}(x_0,\tau)}{N} - \frac{\left[D_E^{(1)}(x_0,\tau)\right]^2}{N}} \quad . \tag{23.7}$$

In sum the statistical error in the estimated drift function is proportional to $(N\tau)^{-1/2}$. Figures 23.1 and 23.2 illustrate the divergent behaviour of the estimated drift coefficient $D_E^{(1)}$ in the cases of few data-points and small time increments considering as example data from an Ornstein-Uhlenbeck process.

23.5 Conclusion

In conclusion there is simple method to estimate the dynamical drift and diffusion functions from measured data. This method can be used to describe the dynamical behaviour of complex systems such as wind energy converters. Quantitative results from this method have to be considered carefully for the reasons discussed in Sects. 23.3 and 23.4.

We would like to stress that there is a more recent extension that avoids the evaluation of the conditional moments in the limit of small time increments and improves the accuracy of the results substantially [4].

References

1. E. Anahua, F. Böttcher, S. Barth, and J. Peinke, Proceedings of the European Wind Energy Conference (EWEC), London, UK (2004)
2. S. Siegert, R. Friedrich, and J. Peinke, Phys Letts A **243**, 275 (1998)
3. H. Risken, *The Fokker-Planck equation*, 2nd ed. (Springer, Berlin Heidelberg New York (1989)
4. D. Kleinhans, R. Friedrich, A. Nawroth, and J. Peinke, Phys Lett A **346**, 42 (2005).

24

Scaling Turbulent Atmospheric Stratification: A Turbulence/Wave Wind Model

S. Lovejoy and D. Schertzer

Summary. Twenty years ago, it was proposed that atmospheric dynamics are scaling and anisotropic over a wide range of scales characterized by an elliptical dimension $D_\mathrm{s} = 23/9$, shortly thereafter we proposed the continuous cascade "fractionally integrated flux" (FIF) model and somewhat later, causal space–time extensions with $D_\mathrm{st} = 29/9$. Although the FIF model is more physically satisfying and has been strikingly empirically confirmed by recent lidar measurements (finding $D_\mathrm{s} = 2.55 \pm 0.02$, $D_\mathrm{st} = 3.21 \pm 0.05$) in classical form, its structures are too localized, it displays no wave-like phenomenology.

We show how to extend the FIF model to account for more realistic wave-like structures. This is achieved by using both localized and unlocalized space–time scale functions. We display numerical simulations which demonstrate the requisite (anisotropic, multifractal) statistical properties as well as wave-like phenomenologies.

24.1 Introduction

According to a growing body of analysis and theory (e.g., [1–5] and references therein) the 1D horizontal sections of the atmosphere follow Kolmogorov laws: $\Delta v = \varepsilon^{1/3} \Delta x^{1/3}$ and (assuming no overall advection/wind): $\Delta v = \varepsilon^{1/2} \Delta t^{1/2}$, where ε is the energy flux. However, 1D vertical sections follow the Bolgiano–Obukov law: $\Delta v = \phi^{1/5} \Delta z^{3/5}$, ϕ the buoyancy variance flux. The generalization to arbitrary space–time displacements $\Delta R = (\Delta r, \Delta t)$, $\Delta r = (\Delta x, \Delta y, \Delta z)$, and multifractal statistics is the 23/9D spatial model, 29/9D space–time model. The anisotropic extension of the Kolmogorov law for the velocity (v) and of the Corrsin–Obukov law for passive scalar density (ρ) satisfy

$$\Delta v\left(\underline{\Delta R}\right) = \varepsilon_{\llbracket\underline{\Delta R}\rrbracket}^{1/3} \llbracket\underline{\Delta R}\rrbracket^{1/3}; \quad \Delta\rho\left(\underline{\Delta R}\right) = \chi_{\llbracket\underline{\Delta R}\rrbracket}^{1/2} \varepsilon_{\llbracket\underline{\Delta R}\rrbracket}^{-1/6} \llbracket\underline{\Delta R}\rrbracket^{1/3}, \qquad (24.1)$$

where ε and χ are the (multifractal) energy and passive scalar variance fluxes; the subscripts indicate the scale at which they are averaged. The key idea of this model is that the physical scale function $\llbracket\underline{\Delta R}\rrbracket$ replaces the usual

136   S. Lovejoy and D. Schertzer

Euclidean distance in the classical isotropic turbulence laws. $\llbracket \Delta R \rrbracket$ need only satisfy the general functional ("scale") equation

$$\llbracket T_\lambda \underline{\Delta R} \rrbracket = \lambda^{-1} \llbracket \underline{\Delta R} \rrbracket; \quad T_\lambda = \lambda^{-G}, \tag{24.2}$$

where T_λ is the scale changing operator which transforms vectors into vectors reduced by factors of λ in scale; T_λ defines a group with generators G. The trace of G is called the "elliptical dimension" D_{el}. The basic 29/9D model has G with eigenvalues 1, 1, H_z, H_t with $H_z = 5/9$, $H_t = 2/3$ so that Trace $G = 29/9$. An overall advection (Gallilean transformation) is taken into account by introducing off-diagonal terms in G. This "generalized scale invariance" (GSI) is the basic framework for defining scale in anisotropic scaling systems [6].

The FIF [7] is a multifractal model obeying (24.2) based on a subgenerator $\gamma_\alpha(\underline{r}, t)$ which is a Levy white noise (index $0 < \alpha \leq 2$; "universal" multifractals [7]). One next obtains the generator $\Gamma(\underline{r}, t)$

$$\Gamma(\underline{r}, t) = \gamma_\alpha(\underline{r}, t) * g_\varepsilon(\underline{r}, t); \quad \widetilde{\Gamma}(\underline{k}, \omega) = \widetilde{\gamma_\alpha}(\underline{k}, \omega)\widetilde{g}_\varepsilon(\underline{k}, \omega) \tag{24.3}$$

by convolving ("*") $\gamma_\alpha(\underline{r}, t)$ with the propagator (space–time Green's function) $g_t(\underline{r}, t)$ where we have indicated fourier transforms by tildes. The conserved flux ε is then obtained by exponentiation

$$\varepsilon(\underline{r}, t) = e^{\Gamma(\underline{r}, t)}. \tag{24.4}$$

The horizontal velocity field is obtained by a final convolution with the (generally different) propagator

$$v(\underline{r}, t) = \varepsilon^{1/3}(\underline{r}, t) * g_v(\underline{r}, t); \quad \tilde{v}(\underline{k}, \omega) = \widetilde{\varepsilon^{1/3}}(\underline{k}, \omega)\widetilde{g}_v(\underline{k}, \omega). \tag{24.5}$$

In order to satisfy the scaling symmetries, it suffices for the propagators to be (causal) powers of scale functions

$$g^{(H)}(\underline{\Delta R}) = h(t)\llbracket \underline{\Delta R} \rrbracket^{-(D-H)}; \quad \tilde{g}^{(H)}(\underline{\Delta R}) = \tilde{h}(t) * \llbracket \underline{k}, \omega \rrbracket^{-H}. \tag{24.6}$$

The real space and fourier scale functions are different; the latter is scaling with respect to the transpose of G, i.e., satisfies (24.2) with G^{T} in place of G. H must be chosen $= D(1 - 1/\alpha)$ for g_ε, and $H = 1/3$ for g_v. $h(t)$ is the Heaviside function necessary to account for causality.

24.2 An Extreme Unlocalized (Wave) Extension

Although the FIF is quite general, its classical implementation ([7,8], Fig. 24.1 upper left) is obtained by using the same localized space–time scale function for both g_ε and g_v. We now allow g_v to be nonlocal in space–time (wave packets). The key is to break $\llbracket \underline{\Delta R} \rrbracket$ into a separate spatial and temporal parts; in a frame with no advection, the classical FIF uses the strongly localized

24 Scaling Turbulent Atmospheric Stratification 137

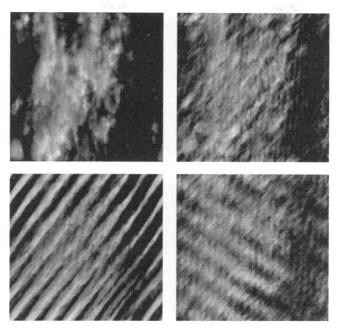

Fig. 24.1. This figure shows a horizontal cross-section of a passive scalar FIF model with varying localization using $g_v(\underline{\Delta R}) = g_{v,\text{wav}}{}^{(H_{\text{wav}})} * g_{v,\text{tur}}^{(H_{\text{tur}})}(\underline{\Delta R})$ and $H_{\text{wav}} + H_{\text{tur}} = H = 1/3$, $H_t = 2/3$; $C_1 = 0.1$, $\alpha = 1.8$ with a small amount of differential anisotropy (using a G with small off-diagonal components). Clockwise from the upper left we have $H_{\text{wav}} = 0, 0.33, 0.52, 0.38$. Rendering is with single scattering radiative transfer. The forcing flux ($\phi = \chi^{3/2}\varepsilon^{-1/2}$) is the same in all cases so that one can see how structures become progressively more and more wave-like. Since we simplify by modeling ϕ directly, the statistics are the same as for a horizontal component of the wind field. See the multifractal explorer: http://www.physics.mcgill.ca/~gang/multifrac/index.htm

$$[\![\underline{\Delta R}]\!]_{\text{tur}} = \left(\|\underline{\Delta r}\|^{H_t} + \Delta t\right)^{1/H_t} ; \quad [\![\underline{k},\omega]\!]_{\text{tur}} \approx \left(\|\underline{k}\|^{H_t} + |\omega|\right)^{1/H_t}. \qquad (24.7)$$

For extreme nonlocalized (wave) models, one can instead choose

$$[\![\underline{k},\omega]\!]_{\text{wav}} = \left(\text{i}\left(\omega - \|\underline{k}\|^{H_t}\right)\right)^{1/H_t}. \qquad (24.8)$$

Taking the inverse Fourier transform of the $-H$ power with respect to ω and ignoring constant factors (see Fig. 24.1 for numerical simulation)

$$\tilde{g}_{v,\text{wav}}^{(H)}(\underline{k},t) = h(t)\, t^{-1+H/H_t} e^{\text{i}\|\underline{k}\|^{H_t} t}. \qquad (24.9)$$

This propagator is a causal temporal integration of order H/H_t of waves.

We have shown that the FIF framework is wide enough to include wave effects. Elsewhere [1], we show with dispersion relations sufficiently close to the standard gravity wave dispersion relations that the model can quite plausibly explain the empirical results in much of the atmospheric gravity wave literature.

References

1. S. Lovejoy, S.D., M. Lilley, et al., Q.J.R. Meteor. Soc. (in press) (2006)
2. M. Lilley, S. Lovejoy, S.D., et al., Q.J.R. Meteor. Soc. (in press) (2006)
3. A. Radkevitch, S. Lovejoy, K.B. Strawbridge, et al., Q.J.R. Meteor. Soc. (in press) (2006)
4. M. Lilley, S. Lovejoy, K. Strawbridge, et al., Phys. Rev. E **70**, 036307 (2004)
5. D. Schertzer and S. Lovejoy, in *Turbulent Shear Flow 4*, edited by B. Launder (Springer, Berlin Heidelberg New York, 1985), p. 7
6. D. Schertzer and S. Lovejoy, Phys.-Chem. Hydrodyn. J. **6**, 623 (1985)
7. D. Schertzer and S. Lovejoy, J. Geophys. Res. **92**, 9693 (1987)
8. D. Marsan, D. Schertzer, and S. Lovejoy, J. Geophy. Res. **31D**, 26 (1996)

25

Wind Farm Power Fluctuations

P. Sørensen, J. Mann, U.S. Paulsen and A. Vesth

Summary. For the simulation of power fluctuations from wind farms, the power spectral density of the wind speed in a single point, and the coherence between two points in the same height but with different horizontal coordinates are estimated based on wind measurements on Risø's test station for large wind turbines in Høvsøre, Denmark.

25.1 Introduction

A major issue in the control and stability of electric power systems is to maintain the balance between generated and consumed power. Because of the fluctuating nature of wind speeds, the increasing use of wind turbines for power generation has caused more focus on the fluctuations in the power production of the wind turbines, especially when the wind turbines are concentrated geographically in large wind farms.

An example of this is observations of the Danish transmission system operator (TSO). Based on measurements of power fluctuations from the 160 MW offshore wind farm Horns Rev in western Denmark, the Danish TSO has found that the fluctuating nature of wind power introduces several challenges to reliable operation of the power system in western Denmark, and also that the wind power contributes to deviations in the planned power exchange with the central European (UCTE) power system, see Akhmatov et al. [1]. It was also observed that the time scale of the power fluctuations was from tens of minutes to several hours.

A model for simulation of simultaneous wind speed fluctuations in many points – corresponding to many wind turbines – with different horizontal coordinates has been developed by Sørensen et al. [2] and implemented in the software program PARKWIND.

The model used in PARKWIND is based on power spectral density (PSD) of the wind speed in a single point, and the coherence between wind speeds in two points. The purpose of the present work has been to assess the presently used PSD and coherence functions. For that purpose, wind measurements on Risø's test site for large wind turbines in Høsøre, Denmark have been used.

25.2 Test Site

The test site in Høvsøre is outlined in Fig. 25.1. It is located only a few km from the west coast of Jutland, on a very flat terrain. The test station is designed to enable test of five large wind turbines simultaneously. For the test of each of the wind turbine, a met mast is erected 247 m west of the wind turbine. Besides, two light masts are erected as indicated in the figure.

The measurements used in the present work are acquired on the met masts. The positioning of these masts on a line is very useful for studies of coherence. Only results from met mast 5 and met mast 4 are presented in this paper. The distance between neighbouring met masts is 307 m, and the line is oriented 4 degrees from straight north–south. The met masts are equipped with a cup anemometer on the top and a wind direction measurement on a boom 2 m lower. The height of each mast has changed slightly a couple of times during the measurement period, because it has been adjusted to the hub height of

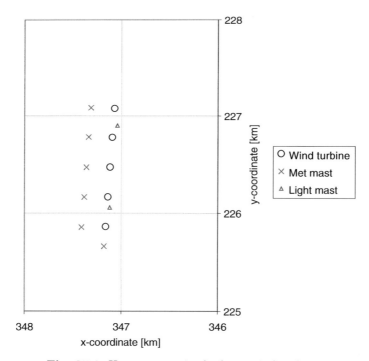

Fig. 25.1. Høvsøre test site for large wind turbines

the wind turbine under test. For the present study, only wind measurements in the height 80±5 m are used.

25.3 PSDs

The PSD of the wind speed in 80 m height is measured, based on data acquired in only 3 months, from 7 July to 9 October 2003. A total of 1787 h data, corresponding to 74 full days, has been acquired and saved in that period. Met mast 5 has been selected for the PSD estimation, and only data in the wind direction sector from 120 to 330 deg has been used to ensure that wake effects from wind turbines are not included.

To obtain values of the PSD for low frequencies, the PSDs are based on time series of 32768 s length, i.e. more than 9 h.

Figure 25.2 shows the resulting measured PSDs for data sectors with 9 h mean wind speed $(7\pm1)\,\mathrm{m\,s^{-1}}$ and $(14\pm2)\,\mathrm{m\,s^{-1}}$, respectively. For comparison, also a Kaimal spectrum [3] is shown. It is seen that the Kaimal spectrum is valid for the higher frequencies, but the measured spectrum has much more energy at low frequencies. To represent this low-frequency energy, a low frequency spectrum has been added, shown as Kaimal + LF.

The Kaimal spectrum is normalised on the well known form

$$n \cdot S_{\mathrm{Kai}}(n) = u_*^2 \cdot \frac{52.5 \cdot n}{(1 + 33 \cdot n)^{5/3}}, \qquad (25.1)$$

where n is a dimensionless frequency, expressed by height above ground z, mean wind speed V_0 and frequency in Hz, f according to (25.2):

$$n = \frac{z}{V_0} \cdot f. \qquad (25.2)$$

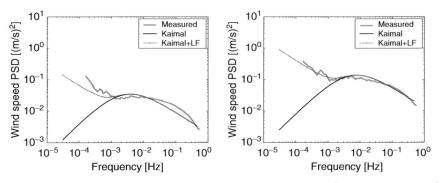

Fig. 25.2. Measured PSD of wind speed time series with $7\,\mathrm{m\,s^{-1}}$ (left) and $14\,\mathrm{m\,s^{-1}}$ (right) mean wind speed compared to Kaimal spectrum and to Kaimal + proposed low frequency spectrum

142   P. Sørensen et al.

A spectrum $S_{\mathrm{LF}}(n)$ for the low frequency fluctuations has been estimated, and a new spectrum $S_{\mathrm{WS}}(n)$ (shown in Fig. 25.2) is defined as the sum of the Kaimal spectrum and the low-frequency spectrum, i.e.

$$n \cdot S_{\mathrm{WS}}(n) = n \cdot S_{\mathrm{Kai}}(n) + n \cdot S_{\mathrm{LF}}(n). \tag{25.3}$$

The low frequency spectrum is estimated based on all measurements from $5~\mathrm{m\,s^{-1}}$ to above. The result of this estimate is given in (25.4).

$$n \cdot S_{\mathrm{LF}}(n) = u_*^2 \cdot \frac{0.0105 \cdot n^{-2/3}}{1 + (125 \cdot n)^2}. \tag{25.4}$$

It is worth noticing in Fig. 25.2 that the new spectrum fits quite well for high frequencies for $7~\mathrm{m\,s^{-1}}$ as well as $14~\mathrm{m\,s^{-1}}$, which indicates that it is reasonable to use all time series with wind speeds above $5~\mathrm{m\,s^{-1}}$ in the estimation of the normalised spectrum, and then apply the estimated normalised spectrum to any mean wind speed.

25.4 Coherence

The coherence analysis presented here is based on all data logged form July 2003 to September 2005. It is necessary to base the coherence analysis on more data than the PSD analysis, because the coherence depends strongly on the wind direction, and therefore only small wind direction sectors can be used.

The Davenport type coherence function [4] between the two points r and c can be defined in the square root form

$$\gamma(f, d_{rc}, V_0) = \mathrm{e}^{-a_{rc} \frac{d_{rc}}{V_0} f}, \tag{25.5}$$

where a_{rc} is the decay factor. Schlez and Infield [5] suggest a decay factor, which depends on the inflow angle α_{rc} shown in Fig. 25.3. The figure shows that $\alpha_{rc} = 0$ corresponds to points separated in the longitudinal direction, whereas $\alpha_{rc} = 90\,\mathrm{deg}$ corresponds to points separated in the lateral direction. With any other inflow angles α_{rc}, the decay factor can be expressed according to

$$a_{rc} = \sqrt{(a_{\mathrm{long}} \cos \alpha_{rc})^2 + (a_{\mathrm{lat}} \sin \alpha_{rc})^2}, \tag{25.6}$$

where a_{long} and a_{lat} are the decay factors for separations in the longitudinal and the lateral directions, respectively. Using our definition of coherence decay factors in (25.5) and (25.6), the recommendation of Schlez and Infield can be rewritten as to use the decay factors

$$a_{\mathrm{long}} = (15 \pm 5) \cdot \frac{\sigma}{V_0}, \tag{25.7}$$

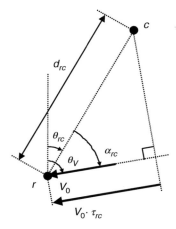

Fig. 25.3. Two points r and c each corresponding to a wind turbine. The distance between the two points r and c is d_{rc}, with a direction θ_{rc} from north. V_0 is the mean wind speed, and θ_V is the wind direction. α_{rc} is the resulting inflow angle, and τ_{rc} is the delay time

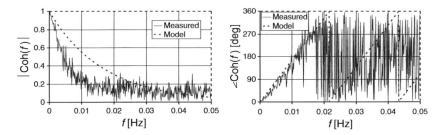

Fig. 25.4. Measured longitudinal coherence between wind speeds on mast 5 and mast 4 based on 2 h segments with mean wind speeds $4\,\mathrm{m\,s^{-1}} < V_0 < 8\,\mathrm{m\,s^{-1}}$

$$a_{\mathrm{lat}} = (17.5 \pm 5)(\mathrm{m\,s^{-1}})^{-1} \cdot \sigma, \tag{25.8}$$

where σ is the standard deviation of the wind speed in $\mathrm{m\,s^{-1}}$.

Figures 25.4 and 25.5 shows measured coherences between wind speeds of mast 5 and mast 4, including only the time series in the wind direction sector from 180 to 190 deg, i.e. longitudinal flow, or $\alpha_{rc} \approx 0$. The result is based on 2 h segments. The coherence is shown for data with 2 h mean wind speeds V_0 in the intervals $4\,\mathrm{m\,s^{-1}} < V_0 < 8\,\mathrm{m\,s^{-1}}$ (Fig. 25.4) and $12\,\mathrm{m\,s^{-1}} < V_0 < 20\,\mathrm{m\,s^{-1}}$ (Fig. 25.5), respectively.

The complex expression of the coherence is used in Fig. 25.4. The left graphs show the amplitude, whereas the right graphs show the phase angle. The measured data is compared to a Davenport coherence (25.5) using Schlez and Infields suggested decay parameter in (25.7). It is observed that the model overestimates the coherence significantly for low wind speeds as well as

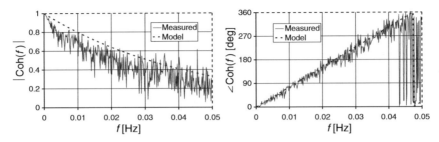

Fig. 25.5. Measured longitudinal coherence between wind speeds on mast 5 and mast 4 based on 2 h segments with mean wind speeds $12\,\mathrm{m\,s^{-1}} < V_0 < 20\,\mathrm{m\,s^{-1}}$

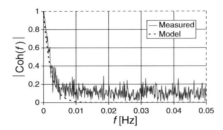

Fig. 25.6. Measured lateral coherence between wind speeds on mast 5 and mast 4 for segments with $4\,\mathrm{m\,s^{-1}} < V_0 < 8\,\mathrm{m\,s^{-1}}$

high wind speeds. The model phase angel is obtained assuming a time delay corresponding to the travel time from r to c with the mean wind speed V_0.

Figure 25.6 shows the measured lateral coherence between wind speeds on mast 5 and mast 4. The lateral direction is here assumed to be with wind directions in the sector from 270 to 280 deg. Comparison to the model show better agreement for lateral decay factor proposed by Schlez and Infield.

25.5 Conclusion

A PSD including the wind speed fluctuations in the time scale from minutes to 9 h has been fitted based on the measurements. The fitted low frequency spectrum applies quite well for different mean wind speeds.

The coherences based on 2 h segments have been compared to coherences proposed by Schlez and Infield, and the conclusion is that the Schlez and Infield coherence overestimates the measured longitudinal coherence, whereas the lateral coherence agrees well with measurements.

References

1. V. Akhmatov, J.P. Kjaergaard, H. Abildgaard. Announcement of the large offshore wind farm Horns Rev B and experience from prior projects in Denmark. European Wind Energy Conference, EWEC 2004. London. November 2004
2. P. Sørensen, A.D. Hansen, P.A.C. Rosas. Wind models for simulation of power fluctuations from wind farms. J. Wind Eng. Ind. Aerodyn. (2002) (No.90), 1381–1402
3. J.C. Kaimal, J.C. Wyngaard, Y. Izumi, O.R. Coté. Spectral characteristics of surface layer turbulence. Q.J.R. Meteorol. Soc. 98 (1972) 563–598
4. A.G.Davenport, The spectrum of horizontal gustiness near the ground in high winds. Quart. J.R. Meteorol. Soc. 87, 194–211
5. W. Schlez, D. Infield. Horizontal, two point coherence for separations greater than the measurement height. Boundary-Layer Meteorology 87. Kluwer, Netherlands 1998. pp. 459–480

26

Network Perspective of Wind-Power Production

Sebastian Jost, Mirko Schäfer and Martin Greiner

26.1 Introduction

Power production in wind farms faces fluctuations on various levels. On the level of a single turbine it is the fluctuation of the wind velocity. The intra-farm wind flow introduces more heterogeneity. At last, the accumulated power output of a wind farm itself represents a volatile source for the power grid. It is this layer which demands for a control of these source fluctuations as well as those resulting from intragrid power redistribution.

New concepts for such a control can be borrowed from modern information and communication technologies. With a distributive routing and conges-tion control self-organizing communication networks are able to adapt to the volatile traffic sources and the current network-wide congestion state [1, 2]. As a result network operation becomes very robust. The exportation of these ideas into wind-powered systems requires to see wind farms as well as the power grid from the network perspective [3].

This contribution represents a first modest step along this network di-rection. It makes use of the critical-infrastructure model proposed in [4], which describes power grids, telecommunication and transport networks in abstracted form. The introduction of a load-dependent metric, a concept which is borrowed from Internet routing, is shown to increase the robustness of such networks against cascades of overload failures. It also reduces respec-tive investment costs. Finally, two model extensions are developed, which are of relevance for a robust interaction between wind-energy sources and the power grid.

26.2 Robustness in a Critical-Infrastructure Network Model

The network model of [4] does not distinguish source, transmitter, and sink nodes. Every node provides/receives flow to/from every other node of the network with an equal share $s_{if} = 1$. After generation at node i, the flow to destination node f is transmitted along the links of the shortest-hop path $[i{\to}f]_{\text{hop}}$. Out of all $N(N-1)$ paths between the N nodes, the betweenness centrality

$$L_n[\text{hop}] = \sum_{i\neq f=1}^{N} \text{path}([i{\to}f]_{\text{hop}};n)s_{if} \qquad (26.1)$$

counts all shortest-hop paths which go over the picked node n and defines its load during normal network operation. The index function $\text{path}([i{\to}f]_{\text{hop}};n)$ is equal to one if n belongs to the shortest-hop path $[i{\to}f]_{\text{hop}}$, and zero else.

In case of a heterogeneous network structure, a very heterogeneous load distribution emerges; see Fig. 26.1. A few nodes have to carry an exceptionally large load. If some of them fail, it comes to a network-wide load redistribution. The shortest flow paths, which were going via the failed nodes, are readjusting and are using the other (transmission) nodes. As a consequence of this readjustment, some nodes have to carry a larger load than before. If the new loads exceed their capacities, then the respective nodes will also fail, triggering a new load redistribution with possibly more overload failure.

In order to reduce the occurrence of such a cascading failure, an $(N{-}1)$ analysis is evoked. One of the N nodes, say m, is virtually removed from the network. The shortest-hop flow paths of the reduced $(N{-}1)$ network are recalculated, which then according to (26.1) determine the readjusted loads $L_n(m)$ of the remaining $N{-}1$ nodes. This procedure is repeated for every

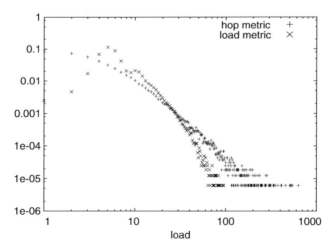

Fig. 26.1. Distribution of node loads L_n resulting from the hop metric (*vertical crosses*) and the load-dependent metric (*rotated crosses*). The two curves are averaged over 50 independent random scale-free network realizations with parameters $N = 1000$ and $\gamma = 3$. A scale-free network is characterized by the degree distribution $p_k \sim k^{-\gamma}$ to find a node attached to k links. The load is given in units of the network size

possible single-node removal $1{\leq}m{\leq}N$. The minimum capacity of node n is then defined as

$$C_n = \max_{0 \leq m \leq N} L_n(m) \,, \tag{26.2}$$

where $L_n(0)$ represents the load resulting from the full network with all N nodes. This assignment guarantees that the network remains robust against a one-node failure, i.e., no overload failures of other nodes and no network-wide cascading failure. A consequence of this gain in network robustness is the increase of investment costs from $\sum_{n=1}^{N} L_n(0)$ to invest $= \sum_{n=1}^{N} C_n$ for the capacity layout.

So far the flow paths have been based on the hop metric. Now a load-dependent metric is introduced. The basic idea is the following: A load-based distance is introduced as

$$d_{i \to f} = \sum_{n=1}^{N} \text{path}(i \to f; n) \, L_n \,. \tag{26.3}$$

The load-based shortest path

$$[i \to f]_{\text{load}} = \arg \left(\min d_{i \to f} \right) \tag{26.4}$$

then minimizes the distance $d_{i \to f}$ between source i and sink f.

For the proper determination of the load-based flow paths the procedure (26.3), (26.4) is not complete. The paths $[i \to f]_{\text{load}}$ are not only determined by the loads L_n of the nodes, the former themselves also determine the latter via

$$L_n[\text{load}] = \sum_{i \neq f = 1}^{N} \text{path}([i \to f]_{\text{load}}; n) \,. \tag{26.5}$$

In order to find a consistent solution of (26.3)–(26.5), these equations have to be treated iteratively.

A first benefit of the load-dependent metric becomes visible in Fig. 26.1. The load-based shortest paths have the tendency to avoid the most-loaded nodes and help to relax the load of the latter. Other nodes which had a small load before acquire a little larger load. As expected, the load-dependent metric turns a heterogeneous load distribution into a more homogeneous one. This sets the stage for a second benefit. Heterogeneous networks with flow paths based on the load-dependent metric are less expensive to establish robustness against one-node-induced cascading failure. Extensive simulations prove that the respective investment cost invest[load] $<$ invest[hop] is smaller than for the hop metric.

A third and maybe largest benefit shows up with the two-nodes-removal analysis. The $(N{-}1)$ analysis does not guarantee network robustness against a simultaneous failure of two or more nodes. A further increase of capacity beyond (26.2) is needed:

150   S. Jost et al.

$$C_n(\alpha) = (1+\alpha)\, C_n \ . \tag{26.6}$$

The tolerance parameter α is assumed to be the same for every node. The larger α, the larger network robustness will be against a two-node-induced cascading failure.

Only a fraction of the nodes is still functioning after such a cascading failure, i.e., their load is still smaller than their capacity (26.6). This fraction does not necessarily form a connected network. Usually these nodes cluster into nonconnected subnetworks. The largest of these subnetworks, i.e. the one containing the largest number N_{gc} of nodes, is called the giant component. Figure 26.2 shows the relative size N_{gc}/N of the surviving giant component, belonging to an initial random scale-free network and obtained after removal of the two most-loaded nodes. As expected, it is close to zero for very small tolerance parameters. For larger values of α the size of the giant component very much depends on the chosen metric. For the load-dependent metric it is significantly larger than for the hop metric. This result is independent of the network size. Consequently, the additional investment costs to establish network robustness beyond one-node failure are also smaller upon application of the load-dependent metric than for the hop metric.

All results presented so far are not restricted to scale-free networks. Other types of networks, like those with a Poisson or exponential degree distribution, have also been studied. Together with a further investigation of a model

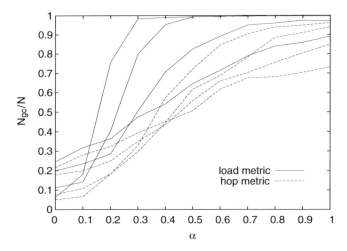

Fig. 26.2. The relative size of the giant network component surviving a cascading failure induced by the removal of the two most-loaded nodes is shown as a function of the tolerance parameter. The *dashed/solid curves* refer to the hop/load-dependent metric, respectively, and have been averaged over 50 independent realizations of random scale-free networks with parameter $\gamma = 3$. From top to bottom at large α, the respective curves correspond to network sizes $N = 1000$, 500, 200, and 100

extension to include link removals and link capacities, the main conclusions from before are confirmed. These are important findings for critical infrastructures like communication networks and power grids.

26.3 Two Wind-Power Related Model Extensions

The critical infrastructure network model of Sect. 26.2 can be extended in many directions. The picture we have in mind for the first generalization is that of a grid consisting of only volatile wind-powered sources. A simple, but adequate model extension is to replace the path strengths s_{if} in (26.1) by random (source) node strengths s_i, which are independently and identically drawn from for example a log-normal distribution

$$p(s) = \frac{1}{\sqrt{2\pi\sigma^2}s} \exp\left\{-\frac{(\ln s + \sigma^2/2)^2}{2\sigma^2}\right\} \qquad (26.7)$$

with fluctuation strength σ. Due to the conserved mean $\langle s \rangle = 1$, the average load $\langle L_n \rangle$ of a node is independent of the fluctuation strength; consult again (26.1). A capacity layout $C_n = (1+\alpha)\langle L_n \rangle$ based on these averaged loads is not able to prevent the occasional overloading of nodes due to the source fluctuations. As a function of the fluctuation strength and for various tolerance parameters, Fig. 26.3a illustrates the relative size of the giant component resulting from a fluctuation-driven cascading overload failure within an initial

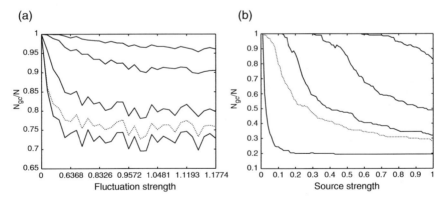

Fig. 26.3. The relative size of the giant network component as a function of (a) the fluctuation strength σ of (26.7) and (b) the source strength s_{wind} of (26.8). All curves are based on the hop metric. From bottom to top the *solid curves* correspond to tolerance parameters $\alpha = 0.1$ (a), 0.0 (b), 0.2, 0.5, 1.0 and the *dotted one* to the $(N-1)$ analysis (26.2). For (a) all curves have been averaged over 50 independent source realizations. One Poisson network realization with degree distribution $p_k = (\lambda^k/k!)e^{-\lambda}$ has been employed; parameters are $N = 100$ (a), 200 (b) and $\lambda = \langle k \rangle = 5$ (a), 7 (b)

152   S. Jost et al.

random Poisson network. For small tolerance parameters and also for the $(N-1)$ analysis based on the average loads, only a relatively small fluctuation strength is needed to knock out a sizeable fraction ($\approx 20\%$) of the network.

The second model extension is motivated from another wind-power related picture: a small fraction $q = 0.1$ of (conventional power plant) nodes with large source capacity s_{cpp} face a large fraction $1-q$ of (repowering wind energy) nodes with small source strength s_{wind}. Instead of (26.7), the source strengths are then assigned according to the bimodal distribution

$$p(s) = q\delta\left(s - rs_{\mathrm{cpp}}\right) + (1-q)\delta\left(s - s_{\mathrm{wind}}\right) . \tag{26.8}$$

In case of $s_{\mathrm{wind}} = 0$, the c.p.p. nodes produce at full capacity s_{cpp} with no reduction ($r = 1$). The setting $s_{\mathrm{cpp}} = 1/q$ then insures $\langle s \rangle = 1$ as in the previous model discussions. With increasing (repowering) s_{wind} the c.p.p. nodes produce less to conserve the overall production ($\langle s \rangle = 1$). This fixes the reduction parameter to $r = (1-q)s_{\mathrm{wind}}$. The capacity layout of the network according to $C_n = (1+\alpha)L_n$ or the $(N-1)$ analysis is done for $s_{\mathrm{wind}} = 0$, which corresponds to the old times when only c.p.p. nodes were produced. With the introduction of source strengths $s_{\mathrm{wind}} \neq 0$, small at first but then further increasing (repowering), again the concern is on the robustness of the network against cascading overload failure. Figure 26.3b reveals that already relatively small source strengths suffice to lead to a major network disruption. Only a further investment into the network in the form of larger tolerance parameters appears to help.

26.4 Outlook

The (toy) model extensions introduced in the previous section allow room for further realistic improvements. Only then it makes sense to import and modify concepts from modern information and communication technologies, like for example those of [1, 2], and to start building a Power Internet. The latter would include the distributive and self-organized monitoring and control of wind farms and the power grid.

References

1. Glauche I., Krause W., Sollacher R., Greiner M. (2004) Phys. A 341:677–701
2. Krause W., Scholz J., Greiner M. (2006) Phys. A 361:707–723
3. Albert R., Barabási L. (2002) Rev. Mod. Phys. 74:47–97
4. Motter A., Lai Y. (2002) Phys. Rev. E 66:065102(R)

27

Phenomenological Response Theory to Predict Power Output

Alexander Rauh, Edgar Anahua, Stephan Barth and Joachim Peinke

27.1 Introduction

This contribution is on power prediction of wind energy converters (WEC) with emphasis on the effect of the delayed response of the WEC to fluctuating winds. Let us consider the wind speed–power diagram, Fig. 27.1a, with a typical power curve of a 2 MW turbine.

Suppose at time t_0 the system is at working point $\{U_0, L_0\}$ outside the power curve with a wind speed U_0 which is constant for a long time. Then, for $t > t_0$, the system will move toward the power curve, either from above or from below. The power curve acts as an attractor [2]. What happens in a fluctuating wind field? Let us follow a short-time series in the wind-speed power diagram, see Fig. 27.1b. We start at some time t_0 with a wind speed $U(t_0)$ and a power output $L(t_0)$. At the next time step, one observes a jump to the point $\{U(t_1), L(t_1)\}$, then to the point $\{U(t_2), L(t_2)\}$, and so on. In Fig. 27.1b, consecutive points are connected by straight lines to form a trajectory. In the ideal case of an instantaneous response of the turbine, and in the absence of noise, all points would lie on the power curve. Actually, the turbine reacts with a delayed response to the wind-speed fluctuations. The timely changes of the wind speed, $\dot{U}$, together with a finite response time hamper accurate power prediction by means of the power curve alone. The application of a suitable reponse theory may help to properly include the influence of turbulent wind in the power assessment. In Fig. 27.2 we show a typical point cluster which is broadly spread around the power curve.

In the following we will discuss in some detail the main idea of a previously published phenomenological response theory [1]. We also will propose an extremal principle to establish an empirical power curve from measurement data [4]. The method is similar but not identical to the attractor principle applied elsewhere [2]. In addition we present an elementary theorem on power prediction in the case of a constant relaxation time of the WEC.

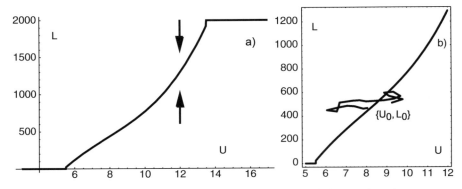

Fig. 27.1. (a) Schematic power curve as an attractor. (b) 20-s (U, L) trajectory in relation to the power curve. Horizontal and vertical units are $\mathrm{m\,s^{-1}}$ and kW, respectively

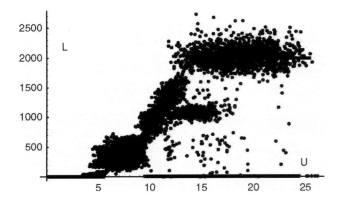

Fig. 27.2. Cluster of 10^4 one Hz points (2 MW turbine at Tjareborg)

27.2 Power Curve from Measurement Data

An inspection of the point cluster in Fig. 27.2 suggests to define an empirical power curve by the location where, in a given speed bin, the maximal density of points $L(t_i)$ is found. This extremal property is expected, if the power curve is an attractor. In previous work [2], the following expectation values were considered $\Delta_{jk} := \langle L(t_{i+1}) - L(t_i) \rangle_{U_j, L_k}$ with the suffix U_j, L_k denoting the restriction to the speed and power bin U_j and L_k, respectively. For a given speed bin U_j, the corresponding point on the power curve was defined by the power bin $L_{k(j)}$ where $\Delta_{jk(j)}$ changes sign. In practice this may cause a problem, if for a given speed bin there are several locations with sign change. The maximum principle, on the other hand, should give a unique result after properly defining the bin sizes:

27 Phenomenological Response Theory to Predict Power Output

$$k(j): \quad N_k := \sum_i L(t_i)|_{L_k, U_j}; \quad N_{k(j)} \geq N_k. \tag{27.1}$$

In words: For a given speed bin j, one determines the number N_k of events in the k-th power bin. The power bin $k(j)$ with the maximal number of events gives the point $\{U_j, L_{k(j)}\}$ of the power curve.

With N_k being the number of events in the k-th power bin, with speed U_j fixed, the statistical error is of the order $\sqrt{N_k}$. After adding these uncertainties to the measured ones $\tilde{N}_k := N_k \pm \sqrt{N_k}$, the intervals $\tilde{N}_k$, possibly, can no longer discriminate between different bins k. In this case one has to increase the bin width and thus the number of events in the bins. The widths of the bins then indicates the likely uncertainty of the curve.

The empirical power curve $L_{\mathrm{PC}}(U)$ as shown in Figs. 27.1 and 27.2 was extracted, by means of the maximum principle, from data of the 2 MW turbine at Tjareborg which were sampled at a rate of 25 Hz over about 24 h [4]. The data were averaged over 1 s which resulted in 87,000 points $\{U(t_i), L(t_i)\}$. 520 1-s data with negative power output and 137 cases with negative wind speed were set to zero, respectively. The width of the speed bins was $1~\mathrm{m\,s^{-1}}$, with values chosen in the middle of the intervals at 3.5, 4.5, 5.5, ... The width of the power bins was variable, in the range from 10 to 50 kW, depending on the number of events. The cut-in and cut-out speeds were chosen at 5.5 and $31.5~\mathrm{m\,s^{-1}}$, respectively, where the latter value was the largest power value of the data set. In the interval $5.5 \leq U \leq 13.5$ the points were fitted by a cubic polynomial. In the interval $13.5 \leq U$, the power was set constant.

Which average power output is predicted by our empirical power curve? To this end, we adopt the standard method by first averaging the data points over 10-min. In Fig. 27.3 such averaged points are plotted corresponding to Fig. 27.2; as should be noticed, Fig. 27.2 depicts the points of a part interval only of length 2.8 h. Next we estimate the power output in two different

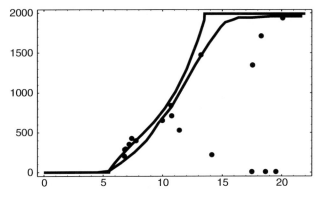

Fig. 27.3. *Dots*: averages of Fig. 27.2 over 10 min without exclusion of shutdown events. Upper and lower curve depict our empirical and the Tjareborg [4] power curve, respectively. Further explanation see text

156   A. Rauh et al.

ways. First the 10-min speeds $\bar{U}_i$ are inserted into the empirical power curve function:

$$< L >_{\mathrm{PC}}= \frac{1}{N_1} \sum_{i=1}^{N_1} L_{\mathrm{PC}}(\bar{U}_i); \quad N_1 = N/600. \tag{27.2}$$

Pictorially, this amounts to shifting the points of Fig. 27.3 vertically onto the power curve. This average is compared with the true average of the measured 1-s powers $L(t_i)$, or equivalently the average of the 10-min values $\bar{L}_i$ ($N_1 = N/600$ be an integer):

$$< L >_{\mathrm{exp}}= \frac{1}{N} \sum_{i=1}^{N} L(t_i) = \frac{1}{N_1} \sum_{i=1}^{N_1} \bar{L}_i. \tag{27.3}$$

An inspection of Fig. 27.3 indicates that L_{PC} significantly overestimates the power output $< L >_{\mathrm{exp}}$, in particular since in the region of the plateau most data points have to be shifted by a relatively large distance from below onto the power curve. As a matter of fact, due to safety reasons, power output is kept limited near the rated power. Also in the large time interval of 24 h our power curve average overestimates $< L >_{\mathrm{exp}}$, i.e., by 17%. In comparison with this, the Tjareborg power curve, available in the world wide Web [4], overestimates the same 24-h data by about 8%.

One reason for this difference may lie in the fact that our 24 h data base for establishing the power curve is rather small. However, in both cases neglection of the finite response time causes systematic errors.

27.3 Relaxation Model

In order to include the delayed reponse of the WEC to power prediction, we recently proposed the following relaxation model [1]:

$$\frac{\mathrm{d}}{\mathrm{d}t} L(t) = r(t) \left[L_{\mathrm{PC}}(U(t)) - L(t) \right]; \quad r(t) > 0, \tag{27.4}$$

where L_{PC} denotes the power curve and $U(t)$, $L(t)$ the instantaneous wind speed and power, respectively. Because the relaxation function $r(t)$ is positive, the above model exhibits the attraction property of the power curve. In principle, the model could be nonlinearly extended by adding uneven powers of $L_{\mathrm{PC}}(U(t)) - L(t)$ with positive coefficients to preserve attraction.

In the simplest case, we may choose $r(t) = r_0 = \mathrm{const}$. Defining the mean power as usual by the time average one finds that

$$< L(t) >=< L_{\mathrm{PC}}(U(t)) > \left[1 + \mathcal{O} \left(\frac{1}{r_0 T} \right) \right]. \tag{27.5}$$

Thus, the average based on the power curve predicts the true mean-power output, provided the averaging time T is much larger than the relaxation time $\tau := 1/r_0$. To see this, one integrates (27.4) from time $t = 0$ to T:

$$\frac{L(T) - L(0)}{Tr_0} = < L_{\mathrm{PC}}(U(t)) > - < L(t) >, \tag{27.6}$$

which implies that the left hand side of the equation tends to zero in the limit of large T.

In reality, a constant relaxation is not observed, see e.g., [3]. In order to implement a frequency-dependent response to wind fluctuations, we made the following linear response ansatz [1]:

$$r(t) = r_0(\bar{U}) + r_1(\dot{U}); \quad r_1(\dot{U}(t)) = \int_{-\infty}^{t} \mathrm{d}t' \, g(t - t') \, \dot{U}(t'). \tag{27.7}$$

Here, r_0 describes relaxation at constant wind speed with $\dot{U}(t) = 0$, compare Fig. 27.1a. The dynamic part $r_1(t)$ of the relaxation function takes into account the delayed response to wind speed fluctuations $\dot{U}(t)$. The function $g(t)$, which simulates the response properties of the turbine, principally may include the control strategies of the turbine at various mean (10-min) wind speeds $\bar{U}$; thus one will generally set $g(t) = g_{\bar{U}}(t)$, see also [3]. In view of the convolution integral in (27.8), one has factorization in frequency space with

$$\hat{r}(f) = \hat{g}(f) \, [2\pi I \, \hat{U}(f)]; \quad I = \sqrt{-1}. \tag{27.8}$$

If the response function $g(t)$ is known together with the power curve, then the average power output can be predicted within this model after an elementary numerical integration of system (27.4) with a given wind field as input. Formally, the dynamic effect can be defined as a correction term, D_{dyn}, to the usual estimate by means of the power curve:

$$< L(t) > = < L_{\mathrm{PC}}(U(t)) > (1 + D_{\mathrm{dyn}}). \tag{27.9}$$

For low turbulence intensities the dynamic correction factor D_{dyn} can be obtained analytically within our model [1]. It has the same structure as the dynamic correction introduced in an ad hoc way in [3]. The comparison with their results [3] allowed us, to deduce the response function $g(t)$ in a unique way, for details see [1]. The response function is shown in Fig. 27.4.

We remark that the ad hoc ansatz made in [3] is limited to small turbulence intensities, whereas our response model can deal, in principle, with arbitrary wind fields.

27.4 Discussion and Conclusion

We have started to derive the response function $g(t)$ from the measurement data [4] by solving (27.4) for $r(t)$ and determining $r(t_i)$ from $L(t_i)$ and $L_{\mathrm{PC}}(U(t_i))$. This has to be done conditioned to different wind speed bins. It turned out that the data base used so far is much too small. There is also the

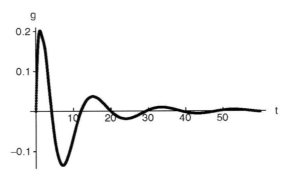

Fig. 27.4. Response function $g(t)$ for the 150 KW turbine Vestas V25 in Beit-Yatir for mean wind speed $\bar{U} = 8$ m s^{-1}, derived in ref. [1] from data of ref. [3]. Time t and g are in units of seconds and reciprocal meter, respectively

problem that, at the Tjareborg site, wind speed is measured relatively far away from the turbine, at a distance of 120 m, which poses the question whether the relevant wind fluctuations acting in the rotor plane are still sufficiently correlated with the wind-speed fluctuations measured far away, especially in view of the response times of the order of a couple of seconds.

In conclusion we have shown, that an empirical power curve can be extracted from high-frequency measurements of only one day. Second, we demonstrated that for a proper power output prediction effects of delayed response have to be considered.

References

1. A. Rauh and J. Peinke, *A phenomenological model for the dynamic response of wind turbines to turbulent wind*, J. Wind Eng. Ind. Aerodyn. 92(2003), 159–183
2. E. Anahua, F. Böttcher, S. Barth, J. Peinke, M. Lange, *Stochastic Analysis of the Power Output for a Wind Turbine*, Proceedings of the European Wind Energy Conference-EWEC in London 2004, as CD
3. A. Rosen and Y. Sheinman, *The average power output of a wind turbine in turbulent wind*, J. Wind Eng. Ind. Aerodyn. 51(1994), 287–302
4. The Tjareborg Wind Turbine Data (1988–1992). Database on Wind Characteristic, http://www.winddata.com, Technical University of Denmark (DTU), Denmark, 2003

28

Turbulence Correction for Power Curves

K. Kaiser, W. Langreder, H. Hohlen and J. Højstrup

Summary. Measured power curves depend on more site-specific parameters than covered by international standards. We present a method to quantify the combined influence of three effects of the turbulence intensity. The method has been tested for different types of turbines with good results. Hence the uncertainty related to site specific turbulence on the power curve can be reduced.

28.1 Introduction

Precise power curve measurements and verifications are important for investors, bankers and insurers. Most contracts specify a procedure according to the IEC regulation [1], to verify the power curve of the purchased WT either on site or under standard conditions. However the IEC regulation does not cover all effects on the measurement relating to site specific wind climates e.g. turbulence. It is known that both anemometers and WTs respond to turbulence in very specific ways. Hence measurements of the power curve will be affected by turbulence, which is related to the topography of the site [2]. The impact of turbulence on the anemometer has been described e.g. in [3]. The qualitative response of WTs to turbulence is known. Due to the complexity of a WT system a quantitative description of the turbine's response is difficult to find. Furthermore the process of binning will introduce an inherent numeric error. Due to the non-linearity, especially near cut-in and rated wind speed, the procedure described in the IEC regulation is sensitive to wind speed variations. These three effects lead to the conclusion that a measured power curve is influenced by the turbulence distribution at the test site [4]. Therefore, power curve measurements of identical turbines at different locations will lead to different results. As a consequence measured power curves have only limited comparable and transferable properties.

28.2 Turbulence and Its Impact on Power Curves

Turbulence intensity depends on a number of site specific conditions and has a wide range of variation. It is related to the standard deviation and can be expressed as follows:

$$I = \frac{\sigma_v}{v} \tag{28.1}$$

with the standard deviation σ of the wind speed and the mean value v.

The numeric error is related to the shape of the ideal power curve. Hence to quantify the numeric error the power curve in homogeneous flow has to be known. In general this information is not available, therefore it is impossible to isolate the numeric error. To eliminate the effects due to turbulence on the power curve, we need to find a method covering the WTs and anemometers response and the numeric error at the same time. Within a wind speed bin the power varies depending on the turbulence intensity [5]. Albers et al. have shown that the mean power in a bin can be described as:

$$\overline{P(v)} = P(\bar{v}) + \frac{1}{2} \frac{\mathrm{d}^2 P(\bar{v})}{\mathrm{d}v^2} \sigma_v^2 \tag{28.2}$$

with $\overline{P(v)}$ mean measured power in a bin $P(\bar{v})$ theoretical power at average wind speed $\bar{v}$ in a bin σ_v standard deviation of the averaged wind speed $\bar{v}$.

From (28.2) it can be seen that the measured power in a bin contains two components, the theoretical power output and a second term, which is related to the second derivative of the power curve and the turbulence. The second term will be negative when the power curve is bent to the right and positive when the power curve is bent to the left. Hence with increasing turbulence intensity the power output will be underestimated at rated wind speed and overestimated near cut-in wind speed (Fig. 28.1). The unknown factors of (28.2) are the ideal power curve at zero turbulence and its second derivative.

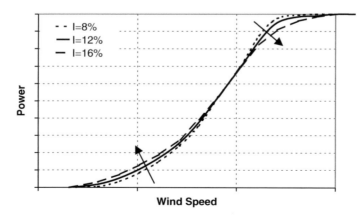

Fig. 28.1. Typical effects of turbulence on power curves

To evaluate a power curve measured with a specific turbulence these two factors have to be determined. Equation (28.1) can be re-written as follows:

$$\overline{P(v)}_i = P_{0_i}(\bar{v}_i) + k_i \sigma_i^2. \tag{28.3}$$

To solve this equation we need for each bin at least two sets of values. If we have more than two sets, the redundant equations can be used to minimise the error of the computation using the least-square method.

28.3 Results

This methodology has been applied for the turbulence intensity correction of power curves for three different turbine types and sites:

- Stall – moderately complex, coastal terrain
- Active Stall – moderately complex, forested terrain
- Variable Speed/Variable Pitch (VS/VP) – flat, inland terrain

Data is being presented in for moderate wind speeds from 7–8 m/s. Figure 28.2 shows three curves each. The horizontal line (dashed) shows the uncorrected

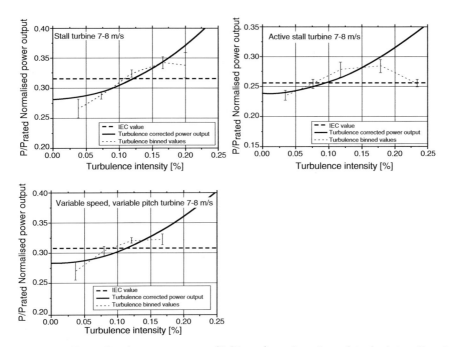

Fig. 28.2. Normalized power output (P/P$_{rated}$) as function of turb. intensity at moderate wind speeds, stall (upper left), active stall (upper right) and VS (bottom) turbines

162   K. Kaiser et al.

mean value according to the IEC standard in the specific bin. The solid curve shows the turbulence corrected value, while the dotted curve shows the data binned in different turbulence intensity intervals. The corrected and the binned data coincide well up to about 12% turbulence intensity. At higher turbulence intensities the binned data deviate from the correction curve. The reason for the deviations is the fact that turbulence fluctuations of the wind speed cover a range down to cut-in wind speed.

28.4 Conclusion

The presented method for correction has been successfully tested on different types of WTs. However, data sets with only marginal variation of turbulence cause problems. The wider the turbulence range within the bins the more reliable the curve fit. As a main advantage of the proposed method it would be possible to transfer measured power curves from one specific site to another with a different turbulence distribution. Furthermore besides standardising power curves with respect to air density a standardisation in relation to turbulence intensity can be introduced.

References

1. IEC 61400-12: Wind Turbine Generator Systems – Part 12: Wind Turbine Power Performance Testing
2. T.F. Pedersen, S. Gjerding, P. Ingham, P. Enevoldsen, J.K. Hansen, P.K. Jorgensen: Wind Turbine Power Performance Verification in Complex Terrain and Wind Farms, Risø-R-1330, Roskilde 2002
3. A. Albers, C. Hinsch: Influence of Different Meteorological Conditions on the Power Performance of Large WECs, DEWI Magazin Nr. 9, 1996
4. K. Kaiser, W. Langreder, H. Hohlen: Turbulence Correction for Power Curves, Proceedings EWEC 2003, Madrid 2003
5. A. Albers, H. Klug, D. Westermann: Outdoor Comparison of Cup Anemometer, Dewi, DEWEK 2000

29

Online Modeling of Wind Farm Power for Performance Surveillance and Optimization

J.J. Trujillo, A. Wessel, I. Waldl and B. Lange

Summary. From practical experience it is known that many reasons for an unsatisfying energy yield of wind farms occur. Particular issues are difficult to discern due to the interaction of the turbines through wake effects. Therefore, reliable wind farm modeling is necessary in order to assess faulty operation and energy production of single turbines. In this paper a method to obtain the expected performance for each single turbine in a wind farm in online basis is presented and tested with real measured data.

29.1 Wind Turbine Power Modeling Approach

Our approach to the single turbine power performance assessment of single turbines is to extend a wind farm model in order to make it work with quasireal time data. Time series of operational data of a wind farm are fed to the model which calculates average power delivered by each single turbine. The modeled power is compared afterward to the measured power data. The wind farm model will be integrated into the wind farm management system OptiFarm^{TM} [1] for wind turbine surveillance purposes.

29.1.1 Wind Farm Model

The selected wind farm model is the farm layout programm (FLaP), an advanced model developed at Oldenburg University [3, 6]. The model manages the superposition of wakes and other relevant wind farm effects to predict an effective wind speed at hub height of each turbine. Finally, the power produced by the turbine is calculated with the power curve of the turbine and the modeled effective wind speed. Extended versions of two wake models, namely Ainslie [2] and Jensen [5] model, have been implemented in FLaP for single wake calculation.

164   J.J. Trujillo et al.

The extended Ainslie wake model assumes a radially symmetric wind speed field around the wake center line. The single wake is modeled with 2D RANS with eddy viscosity closure. An internal turbulence intensity model, implemented by Wessel [7], is used for internal wakes calculation.

Jensen model, defines a wake expanding linearly downwind based on mass conservation. The wake is defined by the expansion angle which is empirically defined by calibration with measured data. In FLaP this model has been implemented making use of two different wake angles that are used for single and multiple wakes, respectively.

29.1.2 Online Wind Farm Model

Extensions to FLaP have been made in order to enable online modeling. Time series of operational parameters are fed to the model performing different checks and corrections to the data as explained later.

Power curve data are corrected for actual air density using the recommendation of the IEC standard [4]. For this purpose data of temperature and pressure have to be given as input for each time series.

FLaP searches the best ambient wind speed measurement available for a particular wind direction giving priority to metmast measurements. The nacelle anemometer, of the active free wind turbine, is used in case of no availability of a free metmast for the actual wind direction. This is made assuming that a proper calibration of the nacelle wind speed is available.

During normal operation of a wind farm there are occasions at which one or more turbines may not operate. This has to be taken into account by the wind farm model since the wake structure inside the wind farm changes. FLaP handles this by simulating with a wind farm layout that only contains the wind turbines that are actually operating at the time for which the simulation is performed.

29.2 Measurements and Simulation

Time series of 10 min averages of measured data were collected at a wind farm located near the North Sea coast of Germany. The wind farm is compounded by thirty four stall regulated turbines with a nominal power of 600 kW in a distribution shown in Fig. 29.1.

Data were available from metmast and SCADA system extending over almost one year. The first system delivered data of wind speed at hub height, wind direction, and air temperature and pressure. The second collected data of power production, nacelle wind speed and status for each turbine.

Simulation of Bassens wind farm time series was performed with FLaP using Ainslie and Jensen models independently.

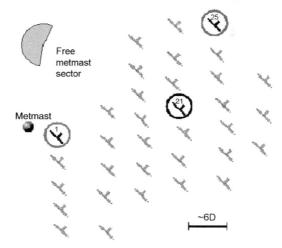

Fig. 29.1. Bassens wind farm layout showing position of metmast and three turbines selected for wind farm model assessment. Free metmast sector between 203° and 15° (in meteorological convention: 0° N, clockwise positive)

29.3 Results

Comparison of measured and modeled power (P_meas and P_mod) is performed by means of mean error and root mean squared error calculated as ME = $\overline{P_{\text{meas},i} - P_{\text{mod},i}}$ and RMSE = $\sqrt{\frac{1}{N} \sum (P'_{\text{meas},i} - P'_{\text{mod},i})^2}$, respectively.

Data were selected for wind directions for which the metmast was receiving free inflow (see Fig. 29.1); enabling study of three different characteristic turbines. Turbine 1 experiences the lowest wake effect, while turbine 21 *observes* wakes from every direction and turbine 25 for some cases is unaffected and *whitstands* the greatest amount of wakes for wind from SW.

In average the wind farm model shows very good agreement with the averaged measured data. As expected, turbines affected by wake during a greater amount of time show higher mean error (see Figs. 29.2 and 29.3). Moreover it can be observed that Ainslie model performs better than Jensen model. However, a tendency to underpredict power for low and overpredict for high wind speeds shows a need in calibration of the thrust coefficient.

The uncertainty for both models is high, especially for low wind speeds, and almost no difference is observed between them. It is to remark that the RMSE presents a dependency with respect to the inflow wind speed and that great part of this uncertainty is related to fluctuations of wind speed and power at each turbine, i.e., to power curve estimation.

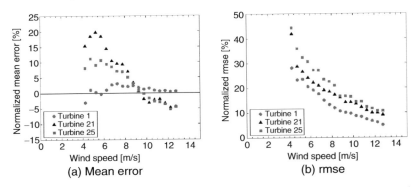

Fig. 29.2. Normalized average mean error and rmse for turbines 1, 21, and 25. Simulation with extended Jensen model. Data selected for free metmast inflow and normalized with respect to power curve values at each wind speed

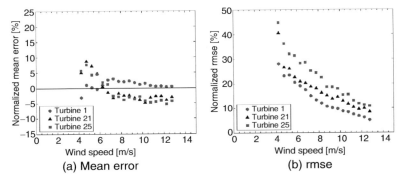

Fig. 29.3. Normalized average mean error and rmse for turbines 1, 21, and 25. Simulation with extended Ainslie model. Data selected for free metmast inflow and normalized with respect to power curve values at each wind speed

References

1. OptiFarm^{TM}: Wind farm management system. www.optifarm.de
2. J.F. Ainslie. Calculating the flowfield in the wake of wind turbines. *J. Wind Eng. Ind. Aerod.* 27:213–224, 1988
3. B. Lange et al. Modelling of offshore wind turbine wakes with the wind farm program FLaP. *Wind Energy*, (1):87–104, January/March 2003
4. IEC. *International Standard IEC 61400–12 Wind Turbine Generator Systems – Part 12: Wind Turbine power performance testing*, 1st edition, 2, 1998
5. N. Jensen. A note on wind generator interaction. Technical Report M–2411, Risø National Laboratory, DK4000 Roskilde, 1983
6. H.P. Waldl. *Modellierung der Leistungsabgabe von Windparks und Optimierung der Aufstellungsgeometrie.* Dissertation, Universität Oldenburg, 175S, 1998
7. A. Wessel. Modelling turbulence intensities inside wind farms. In *EUROMECH 464b Wind Energy*, Carl von Ossietzky University, Oldenburg, 2005

30

Uncertainty of Wind Energy Estimation

T. Weidinger, Á. Kiss, A.Z. Gyöngyösi, K. Krassován and B. Papp

30.1 Introduction

With the development of low cut-in wind speed turbines for continental climatological conditions, wind power has become an economical source of energy in Hungary. The first high power wind turbine was implemented at the end of 2000 (Inota, Balaton Highlands: 250 kW). Currently nine wind turbines are working at a total power of about 7 MW. By the end of the decade 3.6% of the total domestic electric energy production (about 1,600 GW h) is planned to be covered by renewable energy.

Wind-energy research in Hungary has been carried out for more than half a century [1]. Wind maps for the different periods have been prepared [2–4], detailed statistical analysis of the wind data has been done [5] and the Hungarian pages for the European Wind Atlas have been prepared [6,7]. Wind energy maps have been calculated for different levels based on the WAsP model [8] and numerical model results [9].

The purpose of this paper is (1) to analyze the inter-annual variability of wind speed and (2) to demonstrate the uncertainty of the power law formula widely used in the estimation of wind profile.

30.2 Wind Climate of Hungary

The average wind speed in Hungary is between 2 and 4 $\mathrm{m\,s^{-1}}$. The maximum occurs in spring, the minimum in late summer or early fall. Several wind maps of Hungary have similar structure, but in some cases large deviations $(0.2–0.4\ \mathrm{m\,s^{-1}})$ are present in the annual means between maps for different periods or maps based on various interpolation formulae. This feature is demonstrated on two different wind maps for the period from 1997 to 2003 (Fig. 30.1).

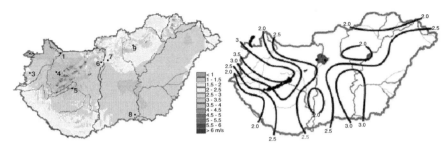

Fig. 30.1. Annual wind speed in Hungary based on 55 automated meteorological stations for the period 1997–2003 [4] (left panel); wind map of Hungary based on wind measurements from 1997 till 2002 [3] (right panel)

Table 30.1. Average wind speed (m s^{-1}) for the period 1970–1987, relative deviation of the maximum and minimum annual mean wind speed from the average (rel. min and rel. max, respectively) and that for wind energy (proportional to the cube of wind speed), estimated from monthly wind roses (cub. min and cub. max, respectively)

Station	Average	rel. min	rel. max	cub. min	cub. max
Győr	2.70 (m s^{-1})	−11.8%	14.3%	−40.3%	31.5%
Sopron	4.38	− 8.9	6.7	−21.6	13.1
Szombathely	3.39	−14.1	14.4	−33.9	82.3
Pápa	2.56	−11.1	8.0	−30.6	26.3
Siófok	3.17	−21.0	23.4	−42.8	58.8
Budaörs	2.88	−14.7	16.5	−34.8	58.9
Budapest	2.87	−15.9	19.9	−46.9	72.7
Szeged	3.32	−6.5	9.6	−14.7	23.5
Kékestető	4.40	−23.1	12.8	−55.2	35.5

Relocation of a weather station also causes significant changes. For example in Sopron (2)[1] the new location of the station is a former wind mill's location; here the annual average wind speed increased by more than 1 m s^{-1}.

The inter-annual variability of annual mean wind speed is an additional source of uncertainty. In Table 30.1 this uncertainty is demonstrated. The calculation is based on Annals of the Hungarian Meteorological Service. The largest values were recorded at the western boundary in Sopron (2) and at the "top" of Hungary, Kékestető (1,010 m elevation). This inter-annual variability is the nature of the wind climate. The deviation of the annual mean from the long term average is around 10%. In the wind tunnel of the Buda hills at Budaörs (6), this type of deviation is about 15%, at Siófok (5) near Lake Balaton or at Kékestető (9) this value is above 20%. For the period of 2001–2004 we got even larger deviations. Annual average wind speed for the latter period

---

[1] Numbers in parenthesis after station names refer to their location in Fig. 30.1.

at Kékestető (9) was only 3.42 m s^{-1}, and at Siófok (5) 3.04 m s^{-1} but contrarily in Pápa (4) it was significantly larger (2.93 m s^{-1}) than the long term average (see Table 30.1 for long term averages of each station).

30.3 The Uncertainty of the Power Law Wind Profile Estimation

Two common methods for the estimation of wind profile include (1) the Monin–Obukhov similarity theory and (2) the power law profile approximation. The similarity theory, however, has large error above the surface layer [10].

The scope of this section is the analysis of the power-law estimation. The independent variables of the power law wind profile formula are the wind speed (U_r) of the reference level (z_r) and the stability dependent exponent (p). The numerical value of the exponent is 1/7 over homogeneous terrain with short vegetation, which is the most common first guess. Dependence on stability (e.g., Pasquill categories) can be assumed through the daily variation of the exponent p. In the daytime convective surface layer its value ranges between 0.07 and 0.1, while by extreme stable stratification $p = 0.3$–0.5 is suggested.

The first derivative of the formula for the wind speed at a certain height ($U(z) = U_r(z/z_r)^p$) with respect to p gives an expression for the sensitivity of the exponential wind profile to p: $\delta U/U = \ln(z/z_r)\delta p$.

In addition, sensitivity for the wind *energy* (which is proportional to the cube of wind speed) is *three times* larger! An error of $\delta p = \pm 0.1$ results in a 20% error at 80 m in the wind speed and 60% in the wind energy! In case of special tower measurements for energy estimation purpose (usually at 10 and 40 m in Hungary) the profile fitting can be done with higher accuracy (as z_0 is higher).

30.4 Inter-Annual Variability of Wind Energy

Extensive wind-energy measurements were initiated in 2001 in Hungary. Let us consider the wind statistics for the period 2001–2003. Year 2001 was especially windy, while year 2003 was calm in Hungary and in the whole Carpathian Basin. The average at Siófok (5) for example in 2001 and 2003 was 3.31 and 2.27 m s^{-1}, respectively. Note that in the data series for Budapest (7) no such significant variability was found, perhaps due to the urban effects.

Let us demonstrate this variability through a practical example! Calculations have been done for a proposed wind farm on the Balaton Highlands.

Continuous wind data were generated from discrete wind measurements (station Szentkirályszabadja, 10) by extrapolating the wind speed in the interval $U_r \pm 0.5$ m s^{-1} for each observation. This continuous data set could be extrapolated smoothly with the power law wind profile to the upper levels.

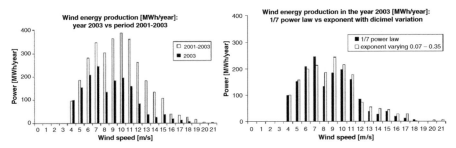

Fig. 30.2. Estimated power of Vestas V90-2.0 wind turbine at 80 m hub height for the period of 2001–2003 vs. the estimated production in year 2003 (left panel) and the estimation made with a stability dependent exponent (right panel)

Estimation for the energy production was done using the power curve of a low cut-in speed 2 MW nominal power equipment (Vestas V90-2.0). Results are demonstrated in Fig. 30.2 (left panel) at 80 m hub height. Compared to the windy year 2001, in 2002 and 2003 we got 18% and 68% less energy, respectively, (see Fig. 30.2 left panel). With the use of a stability dependent exponent instead of the suggested 1/7 one, another 12% uncertainty is introduced into the estimation. Note that the estimation of wind energy was larger with the latter method (see Fig. 30.2).

30.5 Conclusion

The great inter-annual variability is the nature of the wind climate in the Carpathian Basin. The wind energy production of a calm year can be as little as the half of a windy one. Joint analysis of the long-range meteorological data set and wind tower measurements – for at least one year – is fundamental in the estimation of the available wind energy for a certain project.

References

1. Czelnai L. (1953) Időjárás 57:221–227 (in Hung)
2. Péczely Gy. (1981) Climatology (in Hung) Tankönyvkiadó, Budapest
3. Bartholy J., Radics K., Bohoczky F. (2003) Renew Sustain Energy Rev. 7:175–186
4. Wantuchné Dobi I., Konkolyné Bihari Z., Szentimrey T., Szépszó G. (2005) Wind Atlas from Hungary (in Hung) in Wind Energy in Hungary:11–16
5. Radics K., Bartholy J. (2002) Időjárás 106:59–74
6. Dobes H., Kury G. (eds) (1997) Wind atlas Österreichische Beiträge zu Meteorologie und Geophysik 16(378)
7. Bartholy J., Radics K. (2000) Wind Energy Application in the Carpathian Basin. University Meteorological Notes (in Hung) No 14 ELTE, Budapest

8. Tar K., Radics K., Bartholy J., Wantuchné Dobi I. (2005) Magyar Tudomany, 2005.7 (in Hung)
9. Horányi A., Ihász I., Radnóti G. (1996) Időjárás 100:277–301
10. Weidinger T., Pinto J., Horváth L. (2000) Meteorol. Z. 9(3):139–154
11. Irwin J.S. (1979) Atmos. Environ. 13:191–194
12. ISC3 Users Guide (1995) EPA–454/B–95–003a

31

Characterisation of the Power Curve for Wind Turbines by Stochastic Modelling

E. Anahua, S. Barth and J. Peinke

Summary. We investigate how the power curve of a wind turbine is affected by turbulent wind fields. The electrical power output can be separated into two parts namely the relaxation part which describes the dynamic response of the wind turbine on sudden changes in wind velocity and a noise part. We have shown that those two parts describe the power-curve properly if they are calculated from stationary wind measurements. This analysis is very usefull to describe power curve characteristic for situation with increased turbulent intensities and it can be easily applied to measured data.

31.1 Introduction

Let us start with the definition of the power curve which is a nonlinear function of wind velocity

$$L(u) \propto u^3.$$

A standard procedure (IEC 61400-12) to characterise the energy production of a wind electro converter (WEC) is to measure average values of elect. power output and longitudinal wind velocity at the hub hight simultaneously. From those measurements a power curve is obtained by: $\langle u \rangle \to \langle L(u) \rangle$ as shown in Fig. 31.1 (the brackets denote the ensemble averages). This procedure is limited due to (1) nonlinearity of the power curve and (2) relaxation time which describes the dynamic response of the WEC on sudden changes of wind velocity. Such effects lead to the following inequality: $L(\langle u \rangle) \neq \langle L(u) \rangle$. To include those effects into the stationary power curve one could use a Taylor expansion of second-order which is expressed by

$$\langle L(u) \rangle = L(\langle u \rangle) + \frac{1}{2} \frac{\partial^2}{\partial u^2} L(\langle u \rangle) \sigma_u^2 \,,$$

where σ_u^2 is the variance of the wind velocity. Obviously this method is only appropiate for the case of symmetric and weak (quasi-laminar) fluctuations

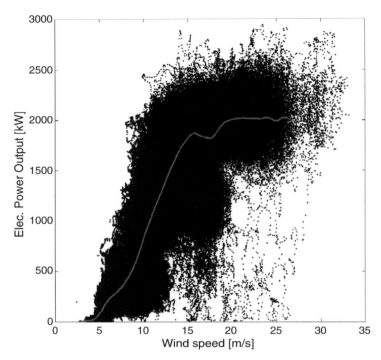

Fig. 31.1. Typical nonlinearity effects on measured power output data of a WEC of 2 MW [5]. A complex behaviour is observed by the dynamics response on sudden changes of wind velocity. The *dots* are instantaneous measured data and the *solid line* is the average power output $\langle L(u) \rangle$

around $\langle u \rangle$, i.e. small turbulent intensities $ti = \sigma_u / \langle u \rangle$. It is known that the wind fluctuation distribution presents an anomalous statistics around the mean value [1], therefore this is again limited. As an improvement, this article reports that the dynamical behaviour of the power output of a WEC, which acts as an attractor, can be described by a simple function of relaxation and noise. We have shown that those two parts describe the power curve properly if they are calculated from stationary wind measurements. This analysis is very usefull to describe power curve characteristics for situation with increased turbulent intensities and it can be easily applied to measured data.

31.2 Simple Relaxation Model

The instantaneous elect. power output of a WEC defined by: $L(t) = L_{\text{fix}}(u) + \ell(t)$ can be described by the following relaxation model [2], see also [6]

$$\ell(t) \propto \text{e}|^{-\alpha t}$$
$$\frac{\text{d}}{\text{d}t} L(t) = -\alpha \ell(t) + g(L,t) \Gamma(t),$$

where $-\alpha\ell(t)$ is the deterministic relaxation of the fluctuations, which growth and decay exponentially, on the stationary power $L_{\text{fix}}(u)$. The term $g(L,t)\Gamma(t)$ describes the influence of dynamical noise from the system, e.g. shutdown states, pitch-angle control, yaw errors, etc. [5].

31.3 Langevin Method

To obtain the stationary power curve by means of *fixed points*, the stochastic temporal state n-vector $L(t) = (L(t_1), L(t_2), ..., L(t_n))$ is assumed to be stationary for wind velocity intervals: $u_a \leq u < u_b$ with an time evolution τ which is described by an one-dimensional Langevin-equation [3,4]

$$\frac{\mathrm{d}}{\mathrm{d}t} L(t) = D^{(1)}(L) + \left(\sqrt{D^{(2)}(L)} \right) \cdot \Gamma(t),$$

where $D^{(1)}$ and $D^{(2)}$ are called *drift* and *diffusion coefficients* and describe the deterministic and stochastic part, respectively. $\sqrt{D^{(2)}}$ describes the amplitude of the dynamical noise with δ-correlated Gaussian distributed white noise $\langle \Gamma(t) \rangle = 0$. These coefficients can be separated and quantified from measured data by the first $(n = 1)$ and second $(n = 2)$ conditional moments [4]

$$M^{(n)}(L, \tau) = \langle [L(t+\tau) - L(t)]^n \rangle |_{L(t)=\mathbf{L}}.$$

Under the condition of $L(u) = \mathbf{L}$. The coefficients are calculated according

$$D^{(n)}(L) = \lim_{\tau \to 0} \frac{1}{\tau} M^{(n)}(L, \tau).$$

Thus, the *fixed points* of the power, $L_{\text{fix}}(u) = \min\{\phi_D\}$, where $\frac{\delta\phi_D}{\delta L} = -D^{(1)}(L)$ is the deterministics potential.

31.4 Data Analysis

The analysis was based on measured data of about 1.6×10^6 samples of elect. power output and wind speed at hub hight of a WEC of 2 MW [5]. First, $D^{(1)}(L)$ was evaluated in a width of the wind velocity bins of $0.5 \, \text{m s}^{-1}$ and power bins of 40 kW. Next, the fixed points of the power were found by searching the $\min\{\phi_D\}$. We show evidence that the power exhibited multiple fixed points for $\langle u \rangle < 20 \, \text{m s}^{-1}$ where the wind generator was switched to other rated speed change by means of a maximal power extraction (optimal operation), e.g. Fig. 31.2. Finally, all fixed points of the power were reconstructed and presented in a two-dimensional vector field analysis $D^{(1)}(L, u)$ of the deterministics dynamics of power and wind velocity, respectively, Fig. 31.3.

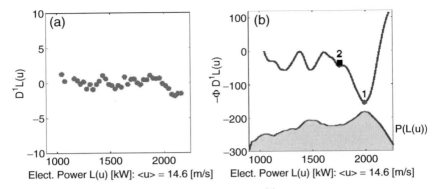

Fig. 31.2. (a) The deterministic dynamics $D^{(1)}(L)$ of the power for $\langle u \rangle = 14.6 \, \text{m s}^{-1}$. (b) The correspondent potential ϕ_D and the power distribution $P(L(u))$. The minima of ϕ_D are the fixed points, stable-states. Position 1 and 2 represents a fixed point and the simple average power, respectively

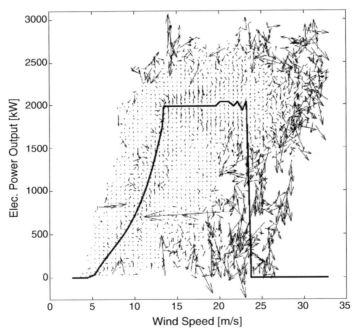

Fig. 31.3. Stationary power curve given by the fixed points for all wind velocity intervals, black line. The arrows represent the deterministic dynamical relaxation of the power output given by a two-dimensional analysis, $D^{(1)}(L, u)$

31.5 Conclusion and Outlook

A simple power output model has been described as a function of relaxation and noise. The stationary power curve has easily been derived by the fixed points method which does not depend any more on the average procedure, i.e. location specific with increased turbulence intensity. Heretofore the relaxation time is described analytically as linear and constant. A detailed analysis on the dynamical relaxation is actually researching together with the response model proposed by Rauh [6] for predicting power output.

References

1. F. Boettcher, Ch. Renner, H.-P. Wald and J. Peinke, (2003) Bound.-Layer Meteorol. 108, 163–173
2. E. Anahua et al. (2004) Proceedings of EWEC conference, London, England
3. H. Risken, (1983) The Fokker-Plank Equation, Springer, Berlin Heidelberg New York, 1983
4. R. Friedrich, S. Siegert, J. Peinke, St. Lueck, M. Siefert, M. Lindemann, J. Raethjen, G. Deuschl and G. Pfister, (2000) Phys. Lett. A 271, 217–222
5. Tjareborg Wind Turbine Data (1992). Database on Wind Characteristic, www.winddata.com, Technical University of Denmark, Denmark
6. A. Rauh and J. Peinke, (2003) J. Wind Eng. Aerodyn. 92, 159–183

32

Handling Systems Driven by Different Noise Sources: Implications for Power Curve Estimations

F. Böttcher, J. Peinke, D. Kleinhans, and R. Friedrich

Summary. A frequent challenge for wind energy applications is to grasp the impacts of turbulent wind speeds properly. Here we present a general new approach to analyze time series – interpreted as a realization of complex dynamical systems – which are spoiled by the simultaneous presence of dynamical noise and measurement noise. It is shown that such noise implications can be quantified solely on the basis of measured times series.

32.1 Power Curve Estimation in a Turbulent Environment

A fundamental problem in wind energy production is a proper estimation of the wind turbine-specific power curve, i.e., the functional relation between the (averaged) wind speed $u(t)$ and the corresponding (averaged) power output $P(t)$: $\langle u(t) \rangle \to \langle P(u(t)) \rangle$. On the basis of this relation and the expected annual wind speed distribution at a specific location the annual energy production (AEP) is estimated.

The main problem of a proper determination of the power curve (officially regularized in *IEC 61400-12*) is its nonlinearity in combination with the turbulent wind field. It is well known that to characterize a fluctuating nonlinear quantity higher order moments are generally needed to be considered. In view of the determination of a proper power curve this means that the association of an averaged power to an averaged wind speed is not unique but will depend at least on the intensity of fluctuations. To circumvent this difficulty it was suggested to expand the power curve into a Taylor series (e.g., [1, 2])

$$P(u) = P_{\mathrm{r}}(V) + \frac{\partial P(V)}{\partial u} \cdot v + \frac{1}{2} \frac{\partial^2 P(V)}{\partial u^2} \cdot v^2 + \mathcal{O}(v^3) \,, \qquad (32.1)$$

where the notation $u(t) = V + v(t)$ (with $V := \langle u(t) \rangle$ and $\langle v(t) \rangle = 0$) for the instantaneous wind speed is used. Assuming that the fluctuations $v(t)$ are

180   F. Böttcher et al.

symmetric around V, neglecting the terms $\mathcal{O}(v^3)$, and averaging (32.1) one obtains:

$$\langle P(u) \rangle = P_{\mathrm{r}}(V) + \frac{1}{2} \frac{\partial^2 P(V)}{\partial u^2} \cdot \sigma^2 , \qquad (32.2)$$

where $\sigma^2 = \langle v^2(t) \rangle$ denotes the variance of fluctuations. Equation (32.2) shows that the "real" power curve $P_{\mathrm{r}}(V)$ (as it would be realized in laminar wind flows) has to be modified. This modification is strong for large variance of fluctuations. For negative curvatures the "real" power curve is underestimated by $\langle P(u) \rangle$ while for positive curvatures it is overestimated.

Obviously (32.2) is only appropriate for small fluctuations, i.e., for small turbulence intensities $\zeta = \sigma/V$. In addition to that $P_{\mathrm{r}}(V)$ is in general not known in advance [2].

We thus propose an alternative approach (based on the theory of Langevin processes) to determine the "real" power curve even for very noisy (turbulent) wind conditions. To this end a simple power curve model will be analyzed.

32.1.1 Reconstruction of a Synthetic Power Curve

A Langevin equation generally describes the temporal evolution of a state vector $\mathbf{q}(t) = (q_1(t), q_2(t), ..., q_n(t))$ and has the following form:

$$\dot{q}_i = D_i^{(1)}(\mathbf{q}) + \sum_j \left(\sqrt{D^{(2)}(\mathbf{q})} \right)_{ij} \Gamma_j(t), \qquad (32.3)$$

where $D^{(1)}$ denotes the deterministic *drift coefficient*, $D^{(2)}$ the stochastic *diffusion coefficient* [3], and $\Gamma(t)$ Gaussian distributed white noise (with $\langle \Gamma_i(t)\Gamma_j(t') \rangle = \delta_{ij}\delta(t-t')$). The coefficients are obtained via the conditional moments

$$M_i^{(1)}(\mathbf{q},\tau) = \langle q_i(t+\tau) - q_i(t) \rangle \,|_{\mathbf{q}(t)=\mathbf{q}},$$
$$M_{ij}^{(2)}(\mathbf{q},\tau) = \langle [q_i(t+\tau) - q_i(t)][q_j(t+\tau) - q_j(t)] \rangle \,|_{\mathbf{q}(t)=\mathbf{q}} \qquad (32.4)$$

by first dividing them by τ and then calculating the limit $\tau \to 0$. For a good temporal resolution the relation between moments and coefficients can be approximated according to

$$M_i^{(1)}(\mathbf{q},\tau) \approx \tau \cdot D_i^{(1)}(\mathbf{q}); \quad M_{ij}^{(2)}(\mathbf{q},\tau) \approx \tau \cdot D_{ij}^{(2)}(\mathbf{q}). \qquad (32.5)$$

In the following we will consider a simple approach identifying $\mathbf{q}(t)$ with $(u(t), P(t))$:

$$\dot{u} = -\alpha\,[u(t) - V] + \sqrt{\beta} \cdot \Gamma(t),$$
$$\dot{P} = \kappa\,[P_{\mathrm{r}}(u(t)) - P(t)] . \qquad (32.6)$$

Thus the only terms entering (32.6) are $D_u^{(1)}$ which is linear in u, $D_{uu}^{(2)}$ as a nonzero constant and $D_P^{(1)}$ which is linear in P. In this case the velocities are

Gaussian distributed around V with variance $\sigma^2 = \beta/2\alpha$ and the difference between the actual and the "real" power decays exponentially. The factor κ (inverse relaxation time) is just a constant in this approach but might well be extended to a more complex function as proposed by [4]. For different mean velocities V and defining

$$P_{\rm r}(u) = \begin{cases} au^3, & u < u_{\rm rated}, \\ P_{\rm rated}, & u \geq u_{\rm rated}, \end{cases} \quad (32.7)$$

the corresponding wind speed and the resulting power output time series can be reconstructed by integration of (32.6). $P_{\rm rated}$ and $u_{\rm rated}$ denote the turbine-specific *rated power* and the corresponding *rated velocity*.

In a first step we apply the *IEC-norm* to determine the power curve from the time series. As can be seen from Fig. 32.1 the resulting power curve differs significantly from the real one, the more the larger the turbulence intensity. This means that although the evolution of the power is purely deterministic the standard averaging procedure is – as a matter of principle – not able to reproduce the correct power curve.

Therefore we suggest in a second step to calculate the drift coefficient according to (32.5) and to search for its zero crossings $D_P^{(1)}(u, P) = 0$ as a new procedure for the evaluation of the "real" power curve. The power curve is reconstructed well by this approach even for very large turbulence intensities such as $\zeta = 30\%$ as also shown in Fig. 32.1.

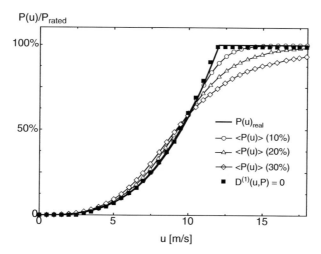

Fig. 32.1. The "real" power curve is indicated by a *thick solid line*. The *open symbols* represent the reconstructed curves according to the *IEC*-standard for turbulence intensities of $\zeta = 10\%$ (*dashed line with circles*), $\zeta = 20\%$ (*dashed line with triangles*), and $\zeta = 30\%$ (*dashed line with diamonds*). The *filled squares* represent the zero crossings of $D_P^{(1)}(u, P) = 0$ for $\zeta = 30\%$

182   F. Böttcher et al.

32.1.2 Additional Noise

So far we assumed a purely deterministic response of the power. On the one side there might be dynamical noise ($D^{(2)} \neq 0$) on the other side measurement noise might be present. In the latter case it is no longer the pure state vector $\mathbf{q}$ that is observed but $\mathbf{q}$ with a superimposed uncertainty:

$$y_i(t) = q_i(t) + \sigma_i \Gamma_i(t). \tag{32.8}$$

In this case the evaluation of the coefficients from the conditional moments according to (32.5) has to be modified [5]:

$$M_i^{(n)}(\mathbf{y}, \tau) = d_i^{(n)}(\mathbf{y}, \tau) + \gamma_i^{(n)}(\mathbf{y}, \tau, \sigma). \tag{32.9}$$

This means that without measurement noise (provided the temporal resolution is good enough) the coefficients are simply given by the slope of the conditional moments as a function of τ. In presence of measurement noise the relation is more complicated. For an Ornstein–Uhlenbeck process the modified slope $d^{(n)}$ and the τ-independent term $\gamma_{(n)}$ can be calculated analytically. For more complex processes they have to be determined numerically.

32.2 Conclusions and Outlook

We presented a new approach to reconstruct the "real" power curve independent of location specific parameters such as the turbulence intensity. So far the analysis is based on a simplified (numerical) wind turbine model but which can easily be extended to more complex systems.

For instance (1) the wind speed can also be modeled by non-Ornstein–Uhlenbeck processes ($D^{(2)} = f(u)$) or non-Langevin processes (in order to get a more realistic spectrum). Additionally (2) the effects of other noise sources can be taken into account according to the approach sketched in (32.9) and finally (3) one might use more wind turbine specific response functions $\kappa = \kappa(u, \dot{u})$.

A first application to real wind turbine systems has been presented by Anahua et al. [6].

References

1. Rosen A., Sheinman Y. (1994) J. Wind Eng. Ind. Aerodyn. 51:287–302
2. Langreder W., Hohlen H., Kaiser K. (2002) Proceedings of Global Windpower Conference, published by EWEA as CD-Rom
3. The symbol $(...)_{ij}$ indicates that first the square root of the diagonalized matrix $D^{(2)}$ has to be taken; successively a back transformation gives the ij-coefficients.
4. Rauh A., Peinke J. (2003) J. Wind Eng. Ind. Aerodyn. 92:159–183
5. Boettcher F. (2005) PhD-thesis, University of Oldenburg
6. Anahua E. et al.(2004) Proceedings of EWEC conference, published on CD-Rom

33

Experimental Researches of Characteristics of Windrotor Models with Vertical Axis of Rotation

Stanislav Dovgy, Vladymyr Kayan and Victor Kochin

Summary. Comparison of the performance characteristics of windrotors with rigidly fixed blades (Dariieu type), and similar windrotors with the control mechanism of blades on the path of their motion, obtained from experimental researches of their models in a hydrotray has shown that the latter possesses the following advantage (a) lower speed of self-start; (b) increase in operating ratio of the flow energy (on 20–70%) and coefficient of rotor torque twice; (c) decrease in some times quantity of full hydrodynamic drag of windrotor models.

33.1 Introduction

Wind is an inexhaustible energy source on Earth. A known converter of wind flow energy into mechanical energy of rotation are windrotors with vertical axis of revolution (Darrieu rotor type). The advantages of wind turbines with such a rotor are independence of their functioning of wind flow direction, transition from console fastening of the windrotor axis to two-basic fastening, a possibility to mount an energy consumer in the basis of wind-plant, simplification of blades design and their fastenings, decrease in aerodynamic noise, etc. One of the main disadvantage of such windrotors is the high velocity of wind flow at which the rotor self-starts rotation. As a result, designers are compelled to supply such wind rotors with additional devices (the electric motor, Savonius rotor, etc.) to spinup the rotor and put it in operating conditions.

The high speed of self-start of a windrotor, if its vertical blades are fixed on horizontal cross-pieces rigidly, is caused by the feature that under static condition a wind flow cannot create sufficient torque to spinup the rotor. If there is a possibility to turn blades of motionless rotor so that the magnitude and direction of aerodynamic force is changed, there will be a possibility that the rotor self-starts even at small wind speed.

33.2 Experimental Installation and Models

In Institute of Hydromechanics NASU, a mechanism has been proposed for controlling the blades position during their movement on a circular trajectory. To research features of operation of a windrotor with such mechanism and compare the results with performance of geometrically similar windrotors with rigidly fixed blades, an experimental setup and several windrotor models with vertical axis have been created.

The windrotors models No. 1 and 2 had different diameter of a circle, on which the axis of blades were placed on the rotor's cross-arms. The number of blades on each model changed during researches from 2 to 4, the blade profile has been chosen axially symmetric of the type NACA-0015.

If blades were fastened on the windrotor cross-arms rigidly (the classical scheme of Darrieu rotor), than the angle of the profile chord of blade to the tangent to rotation circle was equal to $+4°$. The mechanism of blades control provided angular oscillations of blades relative to the blade axis and during one revolution the angle of blade installation changed from -14 to $+25$ grades.

For recording the investigated parameters of rotation of the windrotor model, an automated measuring system of data gathering and processing was used that is a digital storing oscillograph on the basis of the 12-digit 32-channel analog-digital converter L-264 of the firms "L-Card," assembled in the form of an enhancement board of an IBM-compatible computer. The measuring system consisted of three measuring channels.

During experiment the flow velocity $\mathbf{V}_{av}$ in front of the rotor, the revolution velocity "$\mathbf{n}$" of the rotor, and the torque $\mathbf{M}$ on the windrotor model shaft were recorded. The magnitude of torque $\mathbf{M}$ increased until the model stopped.

33.3 Performance Characteristics of Windrotor Models

On the basis of ensemble-averaged values of flow velocity in hydrotray $\mathbf{V}_{av}$, rotational velocity "$\mathbf{n}$" of the model, values of the loading torque $\mathbf{M}$ on the model shaft, and also geometrical parameters of the models, we calculated such windrotor characteristics as the coefficient of specific speed of windrotor $\mathbf{Z} = 2\,\pi\,\mathbf{n}\,\mathbf{R}/\mathbf{V}$, operating ratio of flow energy $\mathbf{C}_p = 2\pi\mathbf{n}\mathbf{M}/\rho\mathbf{V}^3\mathbf{R}\mathbf{l}_{bl}$, and coefficient of rotor torque $\mathbf{C}_m = \mathbf{C}_p/\mathbf{Z}$, where $\mathbf{l}_{bl}$ is the blade length, ρ is density of water.

At constant flow velocity, with increase of the loading torque $\mathbf{M}$ on the model shaft, all rotors have a reduction of rotational velocity "$\mathbf{n}$" of the rotor, but the level of this reduction depends on the rotor design. Rotational velocity decreases faster in rotors with smaller number of blades. Rotors with the mechanism of blade control showed slower speed than rotors with rigidly fixed blades. The reduction of the radius $\mathbf{R}$ of the circle on which the windrotor blades axes move leads to more essential reduction of rotational velocity "$\mathbf{n}$" of the rotor, and consecutively to an increase in the loading torque $\mathbf{M}$ at shaft.

33 Experimental Researches of Characteristics of Windrotor Models 185

It is necessary to note also, that for all considered flow velocities the hydrotray rotors with controlled blades have self-started, while rotors with rigidly fixed blades must be spin upped, i.e., certain auxiliary torque **M** must be applied on the windrotor model shaft. Only after that a rotor continued rotation with constant revolution velocity.

The dependencies C_p on Z for windrotor models with 2, 3, and 4 blades are presented in Fig. 33.1. It appeared, that the positional relationship of curves $C_p(Z)$ for windrotor models with various numbers of blades essentially differ from that of similar curves for windrotor model with rigidly fixed blades. For the last, increasing number of blades leads to a shift of the right part of dome-shaped curve $C_p(Z)$ to the left (i.e., reduces rapidity of a wheel), and on the contrary happens for windrotor models with controlled blades. Here the least high-speed has appeared for the rotor with 2 blades. The greatest values of both the operating ratio of flow energy C_p and the coefficient of windrotor torque C_m ware received just for the windrotor models with 3 and 4 blades.

The influence of the mechanism of windrotor blades control on the magnitude of the torque gained by a windrotor is especially great. The graphs $C_m(Z)$ in Fig. 33.2 show that for both windrotor models with 3 and 4

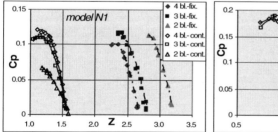
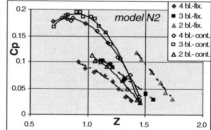

Fig. 33.1. Dependencies of operation ratios of the flow energy C_p on coefficient of specific speed Z (flow velocity in hydrotray $V = 0.6\,\mathrm{m\,s^{-1}}$)

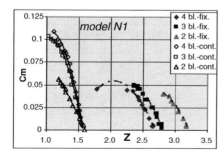
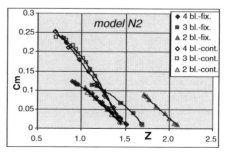

Fig. 33.2. Dependences of coefficients of windrotor model torque C_m on coefficient of specific speed Z (flow velocity in hydrotray $V = 0.6\,\mathrm{m\,s^{-1}}$)

blades the coefficient of windrotor torque C_m at blades control increases 2 times in comparison with windrotor models with rigidly fixed blades. At the same time, the value of coefficient C_m of the windrotor model with 2 blades increases for the model No. 1 only by 30%, and for the model No. 2 by 15–20%.

The measured values of the overall hydrodynamic drag for both models showed that for both windrotor models with blades control a significant decrease in coefficients of hydrodynamic drag was observed, both average Cx_{av}, and maximal Cx_{max}. Thus, for the windrotor model No. 2 coefficients Cx_{av} and Cx_{max} decreased by 30–40%, while for the windrotor model No. 1 those decreased by 3–4 times.

Let us add, that coefficients C_p and C_m were calculated at flow density is equal to $1,000\,kg\,m^{-3}$, that is approximately 750 times the density of air, therefore it will be wrong to compare the results obtained in this work with other known experiments in literature, done in the air environment. From the point of view of advantages of one design of windrotor relative another, we can compare here only our results obtained in absolutely identical conditions of experiment.

33.4 Results

Very high value of the coefficient of rotor torque and low value of self-start speed of windrotor with the investigated mechanism of blades control allow hoping, that wind-plants with such rotors can be rather effective even at rather low speeds of a wind flow as pumping facilities for extraction of water or oil from wells.

34

Methodical Failure Detection in Grid Connected Wind Parks

Detlef Schulz, Kaspar Knorr and Rolf Hanitsch

34.1 Problem Description

This contribution deals with the handling of very common failures and the discussion of outages of wind energy converters (WECs) with doubly-fed induction generators (DFIGs) located in wind parks. Measurements were conducted over a period of time of some months in order to find the tripping source for a high number of WEC-outages. The system overview, a failure detection approach and measurement results are presented. Almost all failures were followed by discussions about the causes for errors and the resulting responsibility due to the high cost of manpower and cost of necessary measurements. One part of the discussion is the sensitivity of the WECs against external over- and under-voltage, transients and voltage dips. These tolerances are given in EN standards and IEC-guidelines [1, 2].

34.2 Doubly-fed Induction Generators

The DFIG's stator is directly connected to the grid. The rotor is connected to a pulse-width modulation (pwm) converter which is connected in turn to the mains. Figure 34.1 shows the system overview of the DFIG. The rotor of a DFIG has, compared to standard machines, a high winding number in order to have a high-rotor voltage at nominal speed. Therefore, no transformer is required on the rotor side. Due to the high-rotor voltage, additional induction effects at stator voltage sags or at fast supply voltage outages may destroy the winding insulation. Transients result, which are avoided by short-circuiting the rotor windings, using the crowbar [3].

As the generator torque change is highly influenced by the control circuit and the reaction of the electrical side to failures, electrical influences contribute to the fatigue loads of the drive train. These effects are not considered for the gear rating, yet. There are different causes for external failures [4, 5]. External distortion sources may be grid-side short-circuits, voltage sags or

188   D. Schulz et al.

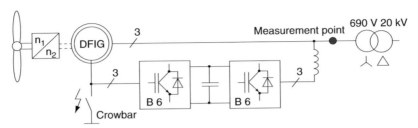

Fig. 34.1. System overview of the DFIG with crowbar and measurement point

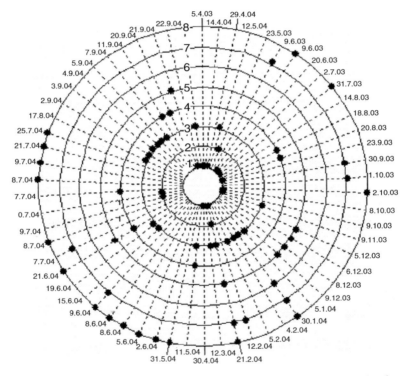

Fig. 34.2. Chronological allocation of WEC-outages of eight systems in the wind park over 18 months, every circle represents one device

transients. Internal sources can be generator short-circuits, dc-link short-circuits or control malfunctions.

34.3 Measurements

Many outages cumulated over some months in the described wind park with eight WECs. The results were: loss of energy yield and additional cost for failure identification. The main point of concern was the determination

34 Methodical Failure Detection in Grid Connected Wind Parks

of the failure source, whether it was an external or device-internal event. Measurements were conducted with an eight-channel power quality analyzer. Figure 34.2 shows a failure-time-circle with the measured tripping events of the eight WECs over a time span of 18 months. With this long-term failure monitoring a methodical evaluation of the failure sources was possible and enabled the operator to plan countermeasures. Our investigations deal with

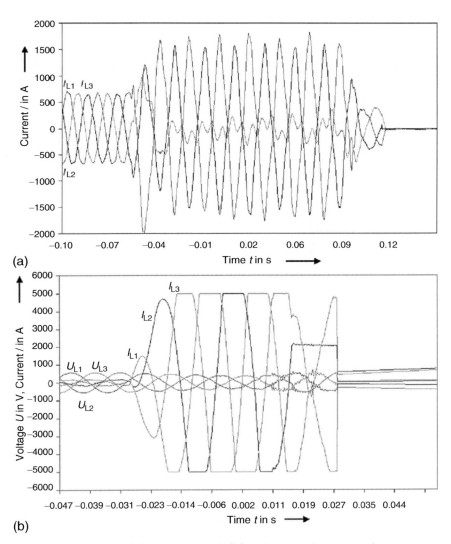

Fig. 34.3. Measured (a) currents and (b) voltage and currents between pwm converter and transformer at internal short-circuit of different 1,5-MW-WEC with DFIG

190   D. Schulz et al.

DFIG outages due to crowbar ignitions. The crowbar closing itself provides no clue about the origin of the failure. Although the failure-time-circle first suggested an external distortion source, dc-link problems were the real cause of the device outages. This reaction was only triggered by some external voltage events far from limits.

During tripping, high voltage peaks on the WEC output were detected by the protection devices. Current increase up to short-circuit values, followed by the shutdown of the WEC were measured on some turbines, see Fig. 34.3a,b. These results show, that the external effects may only trigger an internal device problem.

34.4 Conclusions

The measurements show, that the allocation of failures to internal and external sources causes some effort. This is not always acceptable for wind park operators in practice. Such outages reduce the economic efficiency of wind parks considerably. A detrimental influence on the gear is obvious, but not specified, yet, by simulations or measurements.

References

1. EN 50160 (2000) Voltage Characteristics of electricity supplied by public distribution systems
2. IEC 61000-2-12 (2003) Electromagnetic compatibility (EMC) – Part 2–12: Environment – Compatibility levels for low-frequency conducted disturbances and signalling in public medium-voltage power supply systems
3. Plotkin Y., Saniter C., Schulz D., Hanitsch R. (2005) Transients in doubly-fed induction machines due to supply voltage sags, PCIM Power Quality Conference, November: 342–345
4. Venikow V. (1977) Transient processes in electrical power systems, Moscow, Mir
5. Miri A.M. (2000) Transient processes in electrical power systems (In German: Ausgleichsvorgänge in Elektroenergiesystemen), Springer, Berlin Heidelberg New York

35

Modelling of the Transition Locations on a 30% thick Airfoil with Surface Roughness

Benjamin Hillmer, Yun Sun Chol and Alois Peter Schaffarczyk

Summary. At inboard and mid-span locations of multi-megawatt turbines thick airfoils ($\geq$30%) are increasingly used. The design of the nose area often leads to an increased sensitivity to surface roughness. Experimental research on the airfoil DU97-W-300mod have been carried out including the investigation of discrete roughness by application of spread and discrete roughnesses at Reynolds numbers up to 10 million. These results are compared to simulations with the CFD code FLOWer. The discrete roughness has been modelled both geometrically and by adding a source of turbulent energy.

Key words: Aerodynamics, Transition location, Discrete roughness, Hyperbolic grid generator

35.1 Introduction

The sustainability of wind energy production is considerably affected by electricity production costs per kW h. A reduction of rotor blade costs means in the majority of cases a reduction of masses. Therefore, at inboard and mid-span regions of multi-megawatt turbines thick airfoils ($\geq$30%) are increasingly used. However, the design of the leading edge area results in an increased sensitivity to surface roughness [1, 2]. Multi-megawatt turbines reach Reynolds numbers (Re) from 1×10^6 to 10×10^6 (1–10 M). In order to predict the performance of an installed turbine blade investigations on the effect of surface roughnesses at these Reynolds numbers are important.

At operating turbines surface roughness is mainly caused by smashed insects in the nose region of the blade. It is difficult to model this surface structure exactly. In wind tunnel experiments surface roughnesses are represented by using sandpaper in the nose region of the airfoil as well as by application of trip wire and zigzag tape. The analyses done so far considered mainly the performance depending on Reynolds numbers and on the surface conditions considered as spread roughness [3, 4].

192   B. Hillmer et al.

For a better basic comprehension of the flow regime at discrete roughness and for improved modelling further numerical flow simulations of the airfoil with trip wire and zigzag tape have been carried out. In the simulations the obstacle is aerodynamically represented by modifying the airfoil contour and by modifying the turbulence energy equation.

35.2 Measurements

The numeric results are compared with measurements for the airfoil DU97-300mod with a relative thickness of 30% and a modified trailing edge thickness of 0.49%. The measurements have been carried out in cooperation with the wind turbine manufacturer DeWind at DNW's cryogenetic wind tunnel in Cologne in 2003. Reynolds numbers between 1 and 10 M and Mach numbers (Ma) between 0.1 and 0.2 were obtained by decreasing the fluid temperature down to 100K. Surface roughness was investigated using Carborundum of different grain sizes in the leading edge region as well as zigzag tape and cylindrical trip wire at different locations. In this study experimental results with 11 mm wide zigzag tape of 0.4 mm and 0.6 mm height and trip wire of 1.0 mm height are analysed. The devices are located on the suction side at 30% of the chord.

35.3 Modelling

The used CFD-code FLOWer (release 116.4) has been developed by the German Aerospace Center (DLR) as a part of the research project MEGAFLOW. The code requires a block structured grid and solves the Reynolds averaged Navier–Stokes (RANS) equation. Krumbein [5] implemented a prediction module for the laminar-turbulent transition. The module is coupled to the boundary layer calculation allowing free transition simulations. In the present study an e^N-database method is used as transition criterion. The parameter N is set to $N = 3$. According to Mack's correlation this corresponds to a turbulence intensity of 0.85%. The Wilcox k–ω-model is applied.

Two dimensional C-grids are created with the hyperbolic grid generator KGrid3D [6] developed at the University of Applied Sciences Westküste. The number of cells for the used grid is between 46,000 and 48,000. The airfoil contour line is segmented into 600 cells. In order to achieve a sufficient resolution of the boundary layer calculation the height of the cells at the profile surface is decreased down to 1 μm. In the normal direction to the surface 60–75 cells are created. The wake flow area covers 30 times the chord length using 30 cells in this direction.

The first CFD-model uses a grid created with an obstacle as part of the airfoil contour. Figure 35.1 illustrates the grid around the modelled obstacle.

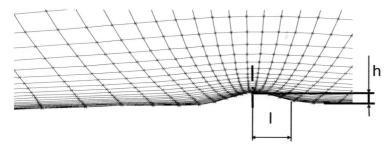

Fig. 35.1. Part of the grid side around the obstacle with height h and length l

The steep slopes of trip wires applicated in wind tunnel experiments could not be modelled. The minimum length l obtained with the used grid generator is five times the modelled obstacle height h.

The second CFD-model keeps the original airfoil contour and attempts to represent the obstacle by adding a turbulent source to the turbulent kinetic energy equation. For this purpose the model from Johansen, Sørensen, Hansen [7] firstly proposed by Hansen [8] was adapted. The two dimensional model originally developed in order to simulate vortex generators is used to compare the effects of different obstacle types. The additional kinetic energy is written as

$$P_k = c_1 \cdot \frac{\rho}{2} \cdot U^3 \cdot \frac{\pi}{2} \cdot h \cdot l \qquad (35.1)$$

with the local flow speed U, the height h and the length l of the discrete roughness in cross-section. c_1 is an empirical constant including the roughness' aerodynamical characteristic regarding the turbulence. The additional kinetic energy is applied at the location of the discrete roughness in an area of $2h \times 2h$.

The simulations are calculated for $Re = 6.6$ M and $Ma = 0.15$.

35.4 Results and Discussion

Figure 35.2 depicts measured and simulated lift coefficients (c_l) against angle of attack. The simulation with the modified airfoil contour with an obstacle height of 0.4–1.0 mm results in lower c_l as the simulation with the turbulence source for a parameter c_1 between 0.25 and 1.

Figure 35.3 shows the measured and simulated c_l against drag coefficients (c_d). The measured c_d exceed the simulation by 0.002 (clean airfoil) to 0.004 (trip wire). The magnitude of the simulated c_d at low angles of attack has been confirmed by xfoil calculations. The difference is perhaps caused by an increased experimental turbulence intensity due to nitrogen injection for cooling purposes. In spite of this discrepancy Fig. 35.3 shows a similar trend in both simulation results and measured values referring to the discrete roughness.

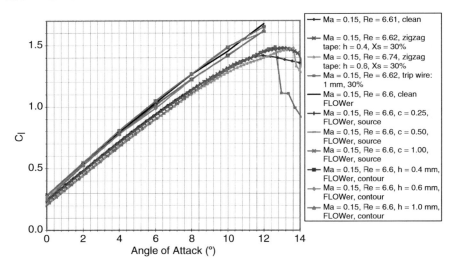

Fig. 35.2. Measured and simulated lift coefficient c_l against the angle of attack α

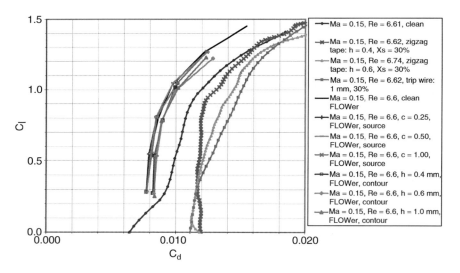

Fig. 35.3. Measured and simulated lift coefficient c_l against drag coefficient c_d

The results of the model with the contour modification exhibit a very small dependency of c_l and c_d on the obstacle height. The results of the model with the variation of the turbulence source exhibit a small dependency of c_l and c_d on the parameter c_1. Separation is not represented sufficiently and the maximum lift coefficient ($c_l^{\max}$) is not correctly predicted. This effect is considered as a shortcoming of the k-ω-turbulence model.

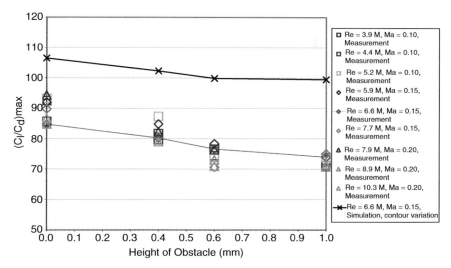

Fig. 35.4. Measured lift/drag ratio, c_l/c_d against roughness height

In Fig. 35.4 the measured lift/drag ratio, c_l/c_d for $3.9\,\text{M} \leq Re \leq 10.3\,\text{M}$ and $0.1 \leq Ma \leq 0.2$ is shown against the obstacle height. A height of 0 means the clean airfoil, a height of 0.4 and 0.6 mm refers to zigzag tape of this height and a height of 1.0 mm refers to the trip wire. Despite a variation due to the range in Mach and Reynolds number and the different type of obstacles the lift/drag ratio clearly exhibits a decreasing trend with the obstacle height. This trend is reproduced by the simulation ($Re = 6.6\,\text{M}$, $Ma = 0.15$) with a contour variation, forming an obstacle of 0.0–1.0 mm of height. Due to an overestimation of c_l and an underestimation of c_d the calculated lift/drag ratio is generally too large. The effect of the modelled additional turbulence source is not shown as the variation of lift/drag ratio is below 1%. Figure 35.5 shows the transition behaviour at an angle of attack $\alpha = 6°$ corresponding to a lift coefficient of $c_l \approx 1$ (design). The effect of the model using a contour variation is seen as a peak. The friction coefficient (c_f) near the peak is decreased. The effect of the additional turbulence source changes the c_f-curve very little. The effect on the transition location is negligible due to earlier natural transition. Further simulations with other locations of the discrete roughness are reasonable.

35.5 Conclusions

A discrete roughness has been modelled with an obstacle as part of the airfoil contour as well as with an additional turbulence source. The simulation results have been compared with measurement data of the airfoil DU97-300mod.

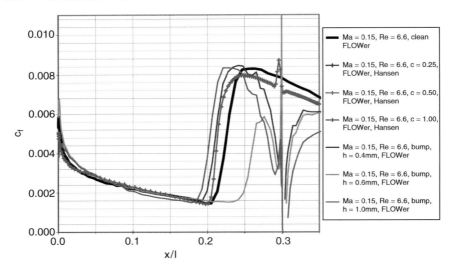

Fig. 35.5. Measured and simulated friction coefficients c_f (suction side) against axial distance over chord x/c at an angle of attack $\alpha = 6°$

The k-ω-model is not able to predict c_l^{max}. The comparison with another turbulence model in future is reasonable. The calculated drag values are considerably below measured values. This is probably due to experimental conditions.

The discrete roughness created with an obstacle as part of the airfoil contour indicates a correct trend with a decreasing lift/drag ratio against the obstacle height. However, the calculated lift/drag ratios are considerably too large. The effect on the transition location is small. The discrete roughness represented by an additional turbulence source does not have a significant effect.

References

1. Rooij R.P.J.O.M. van, Timmer W.A. (2003) Roughness considerations for thick rotor blade airfoils, Journal of Solar Energy Engineering 125:468–478
2. Timmer W.A., Rooij R.P.J.O.M. van (2003) Summary of the delft university wind turbine dedicated airfoils, AIAA-2003-0352:11–21
3. Timmer W.A., Schaffarczyk A.P. (2004) The effect of roughness on the performance of a 30% thick wind turbine airfoil at high Reynolds numbers, Wind Energy 7(4):295–307
4. Freudenreich K., Kaiser K., Schaffarczyk A.P., Winkler H., Stahl B. (2004) Reynolds number and rougness effects on thick airfoils for wind turbines, Wind Engineering, 28(5):529–546
5. Krumbein A. (2002) Coupling the DLR Navier–Stokes Solver FLOWer with an eN-Database Method for laminar-turbulent Transition Prediction on Airoils,

Jahrestagung Strmung mit Ablsung STAB, Numerical Fluid Mechanics – Vol. 77, New Results in Numerical and Experimental Fluid Mechanics III; Siegfried Wagner, Ulrich Rist; Joachim Heinemann; Reinhard Hilbig (eds) Contributions to the 12th AG STAB/DGLR Symposium Stuttgart, Springer, Berlin Heidelberg New York, 2002

6. Trede R. (2003) Entwicklung eines Netzgenerators, Diploma Thesis, FH Westküste, Heide, Germany

7. Johansen J., Sørensen N.N., Hansen M.O.L. (2001) CFD Vortex Generator Model. In: Helm P., Zervos A. (eds) Proceedings of the European Wind Energy Conference, Copenhagen, July 2–6, 2001, pp. 418–421

8. Hansen M.O.L., Westergaard (1995) Phenomenological Model of Vortex Generators. In: Pederson M. (ed) Proceedings of IEA Joint Action, Aerodynamics of Wind Turbines, 9th Symposium, Stockholm, December 11–12, 1995, pp. 1–7

36

Helicopter Aerodynamics with Emphasis Placed on Dynamic Stall

Wolfgang Geissler, Markus Raffel, Guido Dietz, Holger Mai

Summary. Dynamic Stall is a flow phenomenon which occurs on helicopter rotor blades during forward flight mainly on the retreating side of the rotor disc. This phenomenon limits the speed of the helicopter and its manoeuvrability. Strong excursions in drag and pitching moment are typical unfavourable characteristics of the Dynamic Stall process. However compared to the static polar the lift is considerably increased. Looking more into the flow details it is obvious that a strong concentrated vortex, the Dynamic Stall Vortex, is created during the up-stroke motion of the rotor blade starting very close to the blade leading edge. This vortex is growing very fast, is set into motion along the blade upper surface until it lifts off the surface to be shed into the wake. The process of vortex lift off from the surface leads to the excursions in forces and moment mentioned above.

The Dynamic Stall phenomenon does also occur on blades of stall- as well as on commonly used pitch-regulated wind turbines under yawing conditions as well as during gust loads. For any modern wind turbine type Dynamic Stall comes out to be a severe problem.

Time scales occurring during the Dynamic Stall process on wind turbines are comparable to the flow situation on helicopter rotor blades.

In the present paper the different aspects of unsteady flows during the Dynamic Stall process are discussed in some detail. Some possibilities are also pointed out to favourably influence dynamic stall by either static or dynamic flow control devices.

36.1 Introduction

Although a lot of efforts have been undertaken, [1,5,7] the process of Dynamic Stall with all its different flow complexities is still not completely understood nowadays.

The flow is strongly time-dependent in particular during the creation, movement and shedding of the Dynamic Stall Vortex. These rapid flow variations have to be addressed in both numerical as well as experimental investigations. Numerical codes [3], have to be based on the full equations, i.e. the Navier–Stokes equations to solve these problems by taking into account a suitable turbulence model, i.e. [8]. In low speed flows transition plays

200 W. Geissler et al.

an important role as well and has to be taken into account by a transition model, [4]. In helicopter flows the problem of compressibility is present even if the oncoming flow has a Mach number as small as $M = 0.3$. In this case local supersonic bubbles are present during the up-stroke motion. These bubbles are terminated by small but strong shock waves, [2] which trigger the start of the Dynamic Stall process.

36.2 The Phenomenon Dynamic Stall

The advancing helicopter rotor blade encounters the sum of rotation and forward speeds and runs into transonic conditions with the creation of moving shock waves on the blade upper surface. On the retreating blade the difference of rotation and forward speed leads to small local Mach numbers but in order to balance lift on the rotor the incidence has to be increased at this time-instant and Dynamic Stall occurs during this part of the rotor disc.

Figure 36.1 shows the different flow events during Dynamic Stall. The dashed curves indicate the static limit. The lift curve does extend the steady C_{Lmax} by almost a factor of two with an extra peak which is caused by the Dynamic Stall Vortex. After its maximum the lift decreases very rapidly and follows a different path during down-stroke. Of even larger concern is the pitching moment hysteresis loop of Fig. 36.1. It can be shown that the shaded areas between the different parts of the moment loop are a measure of aerodynamic

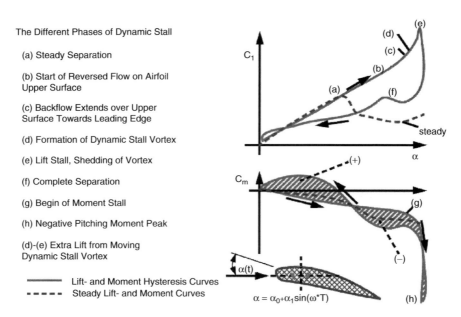

Fig. 36.1. Problem-zones at a helicopter in forward flight

36 Helicopter Aerodynamics with Emphasis Placed on Dynamic Stall 201

Dynamic Stall Vortex During Up-stroke Motion of NACA 0012 Airfoil

Left Figure: $\alpha = 21°$, Right Figure: $\alpha = 25°$
Figures Include Instantaneous Streamlines and Singular Points

Fig. 36.2. Vorticity contours at two time-instants during up-stroke

damping. If these loops are travelled in anti-clockwise sense, the situation is stable, i.e. energy is shifted from the blade into the surrounding fluid. However if the loop is traversed in the clockwise sense (see indication in Fig. 36.2) an unstable flow condition occurs and energy is transferred from the flow to the blade structure. This is a critical situation insofar as dangerous stall flutter may occur.

Figure 36.2 shows the calculated details of the verticals flow at two instants of time during the up-stroke motion. In the left figure the vortex has just been created close to the blade leading edge, has started to travel along the upper surface but is still attached to the surface. In this phase the vortex creates extra lift (see Fig. 36.1). Very short time later (right sequence of Fig. 36.2) the vortex has been lifted off the blade surface, stall has been started and negative vorticity is created at the blade trailing edge moving forward underneath the Dynamic Stall Vortex. In this phase the complex flow phenomenon occur which cause the strong decay of lift and creation of negative aerodynamic damping.

36.3 Numerical and Experimental Results for the Typical Helicopter Airfoil OA209

In October 2004 DLR has done experiments in the DNW-TWG wind tunnel facility located at the DLR-Centre in Göttingen. These tests are part of the DLR/ONERA joint project "Dynamic Stall." Within this project it was decided to use the OA209 (9% thickness) airfoil section as the standard airfoil. The OA209 airfoil is in use on a variety of flying helicopters. Measurements on a 0.3 m chord and 1 m span blade model (extended between wind tunnel side-walls) have been carried out. The objectives of these almost full size tests have been to study the details of the dynamic stall process by both numerical as well as experimental tools. From this knowledge base together with former investigations on this subject possibilities are explored to favourably *control* Dynamic Stall.

Figure 36.3 displays lift, drag and pitching moment hysteresis loops as measured from integrated signals of 45 miniature pressure sensors arranged at mid-span of the blade model. In the plots the force and moment data of all measured 160 consecutive cycles are simply plotted on top of each other. Two sets of data are included:

1. Measurements with straight wind tunnel walls (grey curves)
2. Measurements with steady wind tunnel wall adaptation at mean incidence (black curves).

It is observed that a shift of the lift curve (left upper Fig. 36.3) occurs due to wind tunnel wall interference effects. The results with wind tunnel wall correction do better fit to the numerical curves indicated in the plots as well (dashed curves).

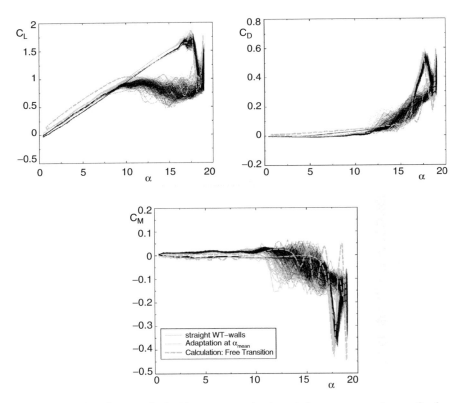

Fig. 36.3. Lift-, drag- and pitching moment hysteresis loops; comparisons of calculation and measurement; $M = 0.31$, $\alpha = 9.8° \pm 9.1°$, $k = 0.05$

36 Helicopter Aerodynamics with Emphasis Placed on Dynamic Stall   203

During the up-stroke phase all measured curves are on top of each other. The flow in this region is non-separated. Close to maximum lift the Dynamic Stall Vortex starts to develop and the different measured curves deviate from each other. This trend continues and is exceeded during the down-stroke phase. The experimental curves show a wide area of distribution until all curves are merging again into a single line at the end of the down-stroke. The numerical curve does fit very close to the experimental results during up-stroke and the first phase of down-stroke. The spreading of the experimental data is caused by turbulence activities at flow separation. This can not be calculated with the present numerical code using turbulence and transition modelling. Nevertheless the correspondence between calculation and measurement in Fig. 36.3 also for drag (upper right Fig. 36.3) and pitching moment (lower Fig. 36.3) is quite satisfactory. All calculated details like surface pressures and skin friction as well as field data (vorticity, density, pressure, etc., not displayed) can be studied and interpreted. Some discussions are included in [6].

36.4 Conclusions

The flow phenomenon Dynamic Stall has been described in some detail and the advantages (high lift) as well as the disadvantage (excursions of drag and pitching moment) have been addressed. Recent experimental data and comparisons with numerical results have been discussed. A good comparison between the 2D-calculations and experimental data has been shown. From this knowledge base the important step to *control* Dynamic Stall in a favourable way is straightforward. Dynamic flow control devices by droop the leading edge of the blade led to considerable improvements. However dynamic control needs actuators strong enough to do the job. The implementation into rotor blades is a formidable task. Therefore passive controlling devices have been considered at DLR and Leading Edge Vortex Generators have been studied. It was found that this type of devices has considerable potential to improve the Dynamic Stall characteristics. Beside of the shape of the leading edge vortex generators their positioning at the blade leading edge is of crucial importance. Due to the fact that the generators may be implemented even on existing blades and that later maintenance problems do not occur, it should be of considerable interest to install the devices also on wind turbine blades to improve its Dynamic Stall characteristics.

In the future considerable effort is planned at DLR both experimentally and numerically to investigate and optimize the effects of leading edge vortex generators.

References

1. Carr L.W. (1985) Progress in Analysis and Prediction of Dynamic Stall, AIAA, Atmospheric Flight Mech Conf Snowmass Co
2. Carr L.W., Chandrasekhara MS (1996) Compressibility Effects on Dynamic Stall. Prog Aerospace Sci Vol 32 pp 523–573
3. Geissler W. (1993) Instationäres Navier-Stokes Verfahren für beschleunigt bewegte Profile mit Ablösung (Unsteady Navier Stokes Code for Accelerated moving Airfoils Including Separation). DLR-FB 92-03
4. Geissler W., Chandrasekhara M.S., Platzer M., Carr L.W. (1997) The Effect of Transition Modelling on the Prediction of Compressible Deep Dynamic Stall, 7th Asian Congress of Fluid Mechanics Chennai (Madras) India
5. Geissler W., Dietz G., Mai H. (2003) Dynamic Stall on a Supercritical Airfoil, 29th European Rotorcraft Forum, Friedrichshafen
6. Geissler W., Dietz G., Mai H., Bosbach J., Richard H. (2005) Dynamic Stall and its Passive Control Investigations on the OA209 Airfoil Section, 31st European Rotorcraft Forum, Florence Italy
7. McCroskey W.J. (1982) Unsteady Airfoils, Ann Rev Fluid Mech Vol 14 pp 285–311
8. Spalart P.R., Allmaras S.R. (1992) A One-Equation Turbulence Model For Aerodynamic Flows, AIAA-Paper 92-0439

37

Determination of Angle of Attack (AOA) for Rotating Blades

Wen Zhong Shen, Martin O.L. Hansen and Jens Nørkær Sørensen

37.1 Introduction

For a 2D airfoil the angle of attack (AOA) is defined as the geometrical angle between the flow direction and the chord. The concept of AOA is widely used in aero-elastic engineering models (i.e. FLEX and HAWC) as an input to tabulated airfoil data that normally are established through a combination of wind tunnel tests and corrections for the effect of Coriolis and centrifugal forces in a rotating boundary layer. For a rotating blade the flow passing by a blade section is bended due to the rotation of the rotor, and the local flow field is influenced by the bound circulation on the blade. As a further complication, 3D effects from tip and root vortices render a precise definition of the AOA difficult.

Today, there exists a lot of experimental and numerical (CFD) data from which precise airfoil data may be extracted. However, this leaves us with the problem of determining lift and drag polars as function of the local AOA. To determine the AOA from e.g. a computed flow field, two techniques have up to now been used. The first technique corresponds to an inverse Blade Element – Momentum (BEM) method in which a measured or computed load distribution is used as input. The technique gives reasonable results and the extracted airfoil data can be tabulated and used for later predictions using the same BEM method. Since the theory is 1D, the accuracy of the extracted airfoil data is restricted to the 1D limitation. Moreover, a BEM code needs a tip correction; different tip corrections results in different airfoil data. The second technique is the averaging technique (AT) employed in [2,3]. This method gives good results, especially in the middle section of a blade. Since the method utilizes averaged data, many input points are needed to evaluate the local flow features. Furthermore, it is difficult to be used for general flow conditions (e.g. at operations in yaw).

In consequence, the goal of the paper is to develop a method that can compute correctly the AOA for wind turbines in standard operations and also in general flow conditions.

37.2 Determination of Angle of Attack

AOA is a 2D concept. For a rotating blade, it can be approached as an AOA in a cross-section. In order to have a uniform inflow in each cross-section, it is then selected in the azimuthal direction (i.e. at constant radius). The selected cross-section is bended in large scale but in small scale can be considered as a plane.

To determine the AOA, we need to do some preliminary work. First, the blade is divided into a few cross-sections (for example $N=10$ sections) where the blade local force is known. Second, a control point should be chosen where the velocity can be read. Theoretically, this control point can be chosen everywhere in the cross-section. In our calculation, the bound circulation around an airfoil section is considered to be a point vortex. Therefore the point should not be too close to the airfoil due to the singularity of point vortex. For convenience, it is chosen in front of the airfoil section. Since the induced velocity varies across the rotor plane, the control point should be chosen in the rotor plane. The features of the choice can be found in Fig. 37.1.

From the initial data, the AOA at each cross-section can be determined. Here we use an iterative procedure to determine the AOA:

Step 1: Determining initial flow angles using the measured velocity at every control point $(V_{z,i}, V_{\theta,i}), i \in [1, N]$

$$\phi_i^0 = \tan^{-1}(V_{z,i}/V_{\theta,i}), \tag{37.1}$$

$$V_{\text{rel},i}^0 = \sqrt{V_{z,i}^2 + V_{\theta,i}^2}. \tag{37.2}$$

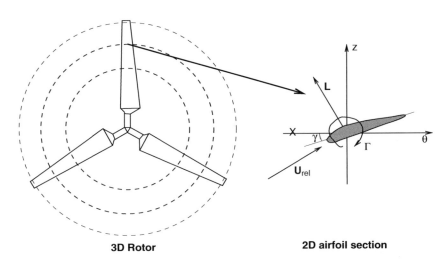

Fig. 37.1. Projection of a 3D rotor in 2D airfoil sections

37 Determination of Angle of Attack (AOA) for Rotating Blades 207

Step 2: Estimating lift and drag forces using the previous angles of attack and the measured local blade forces $(F_{z,i}, F_{\theta,i}), \ i \in [1, N]$

$$L_i^n = F_{z,i} \cos \phi^n - F_{\theta,i} \sin \theta^n, \tag{37.3}$$

$$D_i^n = F_{z,i} \sin \phi^n + F_{\theta,i} \cos \theta^n. \tag{37.4}$$

Step 3: Computing associated circulations from the estimated lift forces as

$$\Gamma_i^n = L_i^n / \left(\rho \sqrt{\Omega^2 r^2 + V_0^2} \right), \tag{37.5}$$

where V_0 denotes wind speed and Ω denotes angular velocity of the rotor. Step 4: Computing the induced velocity created by bound vortex using circulations

$$u_{in}^n(x) = (u_r^n, u_\theta^n, u_z^n) = \frac{1}{4\pi} \sum_{j=1}^{NB} \int_0^R \Gamma^n(y) \times (x - y)/|x - y|^3 dr. \tag{37.6}$$

Step 5: Computing the new flow angle and the new velocity $(V_{z,i} - u_{z,i}^n, V_\theta - u_{\theta,i}^n)$ by subtracting the induced velocity generated by bound vortex

$$\phi_i^{n+1} = \tan^{-1} (V_{z,i} - u_{z,i}^n)/(V_{\theta,i} - u_{\theta,i}^n), \tag{37.7}$$

$$V_{rel,i}^{n+1} = \sqrt{(V_{z,i} - u_{z,i}^n)^2 + (V_{\theta,i} - u_{\theta,i}^n)^2}. \tag{37.8}$$

Step 6: Checking the convergence criteria and deciding whether the procedure needs to go back to Step 2.

When the convergence is reached, the AOA and force coefficients can be determined:

$$\alpha_i = \phi_i - \beta_i, \quad C_{d,i} = 2D_i/\rho V_{rel,i}^2 c_i, \quad C_{l,i} = 2L_i/\rho V_{rel,i}^2 c_i, \tag{37.9}$$

where β_i is the sum of pitch and twist angles and c_i is the chord.

37.3 Numerical Results and Comparisons

In this section, the new developed method is used to determine the AOA for flows past the Tellus 95 kW wind turbine. The Tellus rotor is equipped with three LM8.2 blades rotating at $\omega = 5.0161 \ \mathrm{rad\,s^{-1}}$. Navier–Stokes computations were previous carried out for wind speeds of 5, 7, 9, 10, 12, 15, 17 and 20 m s^{-1} with the EllipSys3D code. For more details about CFD computations, the reader is referred to [2]. Local force on the blade and local velocity at different positions in the rotor plane are then extracted from the data. In Fig. 37.2, drag and lift coefficients at radial position $r = 65.7\%R$ are obtained

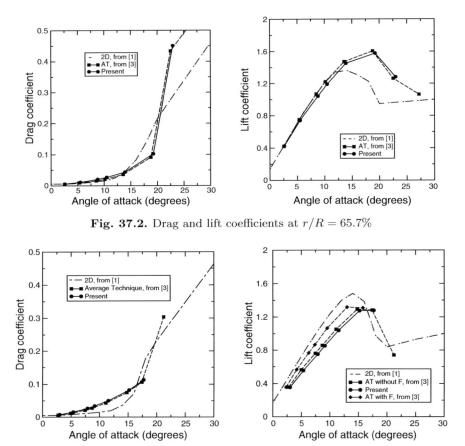

Fig. 37.2. Drag and lift coefficients at $r/R = 65.7\%$

Fig. 37.3. Drag and lift coefficients at $r/R = 94.7\%$

and compared to 2D airfoil characteristics from [1] and the results of AT [2]. Excellent agreement is seen in the linear region up to $\alpha = 12$. For big AOA ($> 15°$), 2D airfoil data fails due to the Coriolis and centrifugal forces present in the boundary layer of a rotating blade. Next we consider the position closer to the blade tip at $r = 94.7\%R$. The drag and lift coefficients are plotted in Fig. 37.3. Since the position is very close to the tip, a tip correction using Prandtl's tip loss function for AT is also plotted. From the figure, we can see that the present result agrees well with the AT without tip correction. On the other hand, the introduction of a tip correction gives the curve closer to the 2D characteristics.

As a summary, the drag and lift coefficients in different radial positions are plotted in Fig. 37.4. From the figure, we can see that drag coefficient increases when the radial position is close to the rotor center whereas lift coefficient remains the same but with a small stall region.

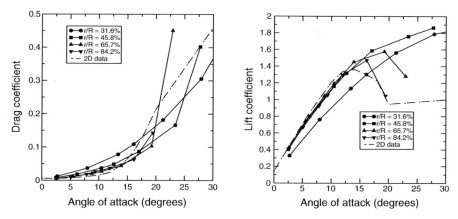

Fig. 37.4. Drag and lift coefficients in function of AOA

37.4 Conclusion

We have now developed a general method for determining AOA. Comparisons with the AT shows good agreements. The method can also be used to extract data from experiments (for example, with Pitot tube). Moreover it can be used to study airfoil properties in the region near the tip.

References

1. Abbott I.H., von Doenhoff A.E. (1959) Theory of wing sections. Dover, New York
2. Hansen M.O.L., Johansen J. (2004) Tip studies using CFD and computation with tip loss models. Wind Energy 7:343–356
3. Johansen J., Sørensen N.N. (2004) Airfoil characteristics from 3D CFD rotor computations. Wind Energy 7:283–294

38

Unsteady Characteristics of Flow Around an Airfoil at High Angles of Attack and Low Reynolds Numbers

Hui Guo, Hongxing Yang, Yu Zhou and David Wood

Summary. An experimental study on unsteady aerodynamic characteristics of NACA0012 airfoils at angles of attack α from $0°$ to $80°$ and Reynolds number of 5×10^3 was carried out in a low speed water tunnel, concerning the starting problems of a small wind turbine. The unsteady characteristics of lift together with PIV and laser sheet visualization results shows that the second maximum lift at $\alpha \approx 40°$ is induced by the large vortex developed from the amalgamation of the smaller vortices from the leading edge.

38.1 Introduction

For a small wind turbine generator, the starting performance has a great influence on the power generating capacity. During the starting process, the blades usually operate at high angles of attack and low Reynolds numbers. The aerodynamic study on this region is essential for improving its starting performance [1]. However, such study was seldom conducted before. Most of the researches concentrated on the problems at low angles of attack and high Reynolds numbers for aeronautics. The flow physics at high angle of attack and low Reynolds number is also hardly to study effectively due to its complexity and lack of theory and advanced test techniques. With the development of wind turbine generators, efforts have been made to address this problem [2–4], but there are still many problems urgently to be solved.

This paper is to report what we have done in this area for the unsteady aerodynamic characteristics of high α and low Re, which can help us to learn more about the flow physics, laying the foundation for understanding and improving the starting performance of small wind turbines.

38.2 Test Facility and Setup

The experiment was conducted in the water tunnel LW-2330 at The Hong Kong Polytechnic University. The test section is $0.3\,\mathrm{m}$ (width) $\times 0.6\,\mathrm{m}$ (height) $\times 2.0\,\mathrm{m}$ (length). The test velocity is $0.05\,\mathrm{m\,s^{-1}}$. The NACA0012 airfoil with

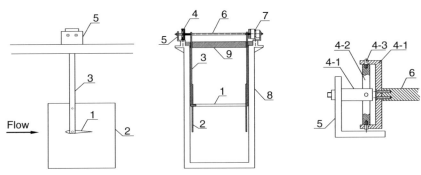

(a) Sketch of side view (b) Sketch of front view (c) Torque resisting system

Fig. 38.1. The test setup: 1 – airfoil, 2 – side plates, 3 – airfoil supporters, 4 – torque resisting system, 5 – setup base, 6 – connection pole, 7 – force sensor system, 8 – water tunnel, 9 – cover plate, and 4-1 – U shaped connectors, 4-2 – circular plate, 4-3 – pins around which the connectors can turn freely

chord length of $c = 0.1$ m is selected as the test model, and the test Reynolds number in terms of the airfoil chord-length is $Re = 5 \times 10^3$. The test angle of attack of the airfoil ranges from 0° to 80°.

A 3-component force sensor, 9251A, made by Kistler Company, was used to measure the instantaneous lift and drag simultaneously. The test setup is shown in Fig. 38.1. The force sensor had to be positioned out of the test section due to its non-water resistant. A unique torque resisting system, shown in Fig. 38.1c, was designed to successfully avoid the influence of any torque. Two side-plates are used to prevent 3-D flow effects.

Particle image velocimetry (PIV) and laser sheet visualization were also used to obtain quantitative information and detailed flow phenomena.

38.3 Experimental Results and Discussions

The typical power spectrum density (PSD) results of the lift at $\alpha = 20°$ and 40° and the corresponding PIV results are presented in Fig. 38.2a,b, respectively, where γ is the circulation Γ around the PIV view field divided by its area. The PSD results of lift will reflect its response to the flow field unsteady properties which are shown by the temporal changes of γ. From the comparison, following findings are summarized:

1. At lower angles of attack, $\alpha \leq 10°$, there is no evident peak in the PSD curve of lift except at $f^* = 1.2$, the self-vibration frequency of model.
2. At $\alpha = 20°$, a peak appears at $f^* = 0.16$ Hz in the PSD curve of lift measurements, and grows to its maximum with the angle of attack increasing to $\alpha = 40°$, when γ nearly has the same primary frequency of $f^* = 0.18$. Such synchronization implies that the second peak lift

38 Unsteady Characteristics of Flow Around an Airfoil 213

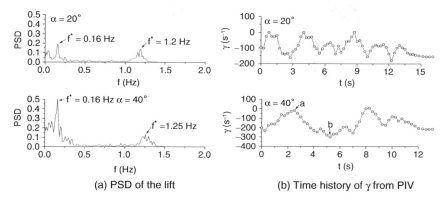

(a) PSD of the lift   (b) Time history of γ from PIV

Fig. 38.2. Unsteady characteristics of (**a**) the lift and (**b**) the corresponding flow field. The flow field unsteadiness is presented in terms of the time history of γ

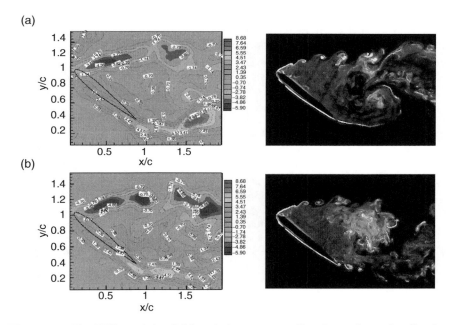

Fig. 38.3. The PIV vorticity field and the corresponding laser-sheet visualization pictures around the airfoil at instant marked "a" and "b" in Fig. 38.2b at $\alpha = 40°$

occurring at $\alpha = 40°$ [3] should be caused by the primary periodic movement in the flow field.
3. For $\alpha \geq 50°$, although flow field is much more periodic, no perceptible influence can be found in the corresponding PSD of force.

Figure 38.3 presents the vorticity field results of PIV and the corresponding laser-sheet visualization pictures of the airfoil at different instant marked "a"

214   H. Guo et al.

and "b" in Fig. 38.2b at $\alpha = 40°$, which correspond to the minimum and maximum lift, respectively. It can be observed from these pictures that:

1. The flow around the airfoil is dominated by alternately rolling up and shedding of large vortices from leading and trailing edge. Laser sheet visualization reveals the separated shear layer rolling up into smaller vortices, smaller vortices amalgamating into a large vortex.
2. Maximum lift occurs when the large vortex just formed from the leading edge and before its shedding downstream, as shown in Fig. 38.3b. The area occupied by this vortex is characterized by higher negative vorticity. When the minimum of the lift is reached, the large vortex evolves from the trailing edge, as shown in Fig. 38.3a.

38.4 Conclusions

This experimental study revealed that the second maximum lift at $\alpha \approx 40°$ is induced by the large vortex developed from the amalgamation of the smaller vortices from the leading edge.

References

1. Wright A.K., Wood D.H., 2004, The starting and low wind speed behavior of a small horizontal axis wind turbine, Journal of Wind Engineering and Industrial Aerodynamics, 92: 1265–1279
2. Wu J.Z., Lu X.Y., Denny A.G., Fan M., Wu J.M., 1998, Post-stall flow control on an airfoil by local unsteady forcing. Journal of Fluid Mechanics 371: 21–58
3. Michos A., Bergeles G., Athanassiadis N., 1983, Aerodynamic Characteristics of NACA 0012 Airfoil in Relation to Wind Generators, Wind Engineering, 7: 247–262
4. Tangler J.L., 2004, Insight into a wind turbine stall and post-stall aerodynamics, Wind Energy, 7(3): 247–260

39

Aerodynamic Multi-Criteria Shape Optimization of VAWT Blade Profile by Viscous Approach

Rémi Bourguet, Guillaume Martinat, Gilles Harran and Marianna Braza

39.1 Introduction

The purpose of this study is to introduce an original blade profile optimization method in wind energy aerodynamic context. The first 2D implementation focuses on Darrieus vertical axis wind turbines (VAWT). This class of VAWT is generally associated with classical aeronautic blade profiles (NACA0012 to NACA0018) whereas Reynolds number and stall conditions are different in aerogeneration case [1]. The objectives of the multi-criteria optimization proposed are to increase the nominal power production, to widen the efficiency range in order to take into account of wind instability and to reduce blade weight and by this way minimizing inertial constraint and mechanical fatigue.

39.2 Physical Model

The two main objective functions of this multi-criteria optimization are related to aerogenerator performances, which means that it is necessary to evaluate these ones for each profile considered.

39.2.1 Templin Method for Efficiency Graphe Computation

In the context of this first study, power production efficiency is computed by using a static method introduced by Templin [2] and which consists in a summation of the global aerodynamic forces applying on the blades during a revolution of the turbine, considering the polar curve of the profile.

39.2.2 Flow Simulation

The flow is simulated by an unsteady Reynolds Average Navier–Stokes approach which consists in splitting every physical variable into two parts:

216   R. Bourguet et al.

a statistical average and a chaotic value. This decomposition applied to Navier–Stokes system leads to an open system where Reynolds tensor introduces six new unknowns. The previous system coupled with a two-equation k-ε closure scheme enables an efficient simulation of the flow, especially with Chen–Kim modification [3] that introduces sensitivity on the turbulent dissipation production term. This physical model is computed on a dedicated auto-generated "C" type meshgrid (50,000 cells) where the profile is fixed and inlet Neumann conditions are variable in order to simulate the evolution of the flow incidence around the blade during a revolution. StarCD (AdapCO, *www.cd-adapco.com*) was selected as CFD code and the numerical scheme used is a PISO predictor/corrector one.

39.3 Blade Profile Optimization

The optimization approach used is a parametric design of experiments (DOE)/response surface models (RSMs) method implemented in Optimus (Noesis, *www. noesissolutions.com*). The first step consists in computing a reduced physical model based on a specific shape parametrization. Blade profiles are parameterized by a 7-control-point Bézier curve built on Bernstein function basis, and thus, the shape parameters are the coordinates of the control points. Solidity is fixed to 0.2. Blade profile chords are constant. As a consequence, only control point ordinates are variable. Moreover, this first study focuses on symmetric profiles, which means that chord-symmetrical control points are coupled. The use of Bézier curve implies that each control point displacement changes the whole profile curve.

39.3.1 Optimization Method: DOE/RSM

The objective of the DOE is to sample each design parameter in order to well describe the shape design space. Taking into account of the rather low number of control points, a factorial approach is convenient. For each point of the DOE, the three cost functions are evaluated:

- *Nominal power production* is the maximum of the efficiency graphe
- *Efficiency range/robustness* is measured as the area under the power curve,
- *Blade weight* is evaluated as the profile area

The results are approximated by a minimization of the quadratic error between the points evaluated and the RSM for each objective function. This least square approximation problem consists in solving a linear system for each cost function. In order to compute a multi-criteria optimization, an hybrid cost function surface is obtained by an equally weighted summation of the RSM associated to each initial criterion.

39.3.2 Reaching the Global Optimum

On the hyper-surface computed by the RSM method, classical optimization algorithms are applied to reach the global optimum. The main deterministic method used in `Optimus` is the Sequential Quadratic Programming approach. It consists in formulating the continuous optimization problem in term of Lagrangian and in solving the Karush–Kuhn–Tucker conditions at each iteration [4]. In case of several local optima, a stochastic method is preferable. The simulated annealing is a very simple one which consists in a random research where a less fitness of the parameters is not always rejected, but with a certain probability. On the other hand, in case of highly irregular RSM, Genetic Algorithms are very efficient to reach the global optimum but imply a higher number of cost function evaluations. The optimization approach has to be chosen considering RSM shape and, in a general case, the last evolutionary method is always convenient.

39.4 Numerical Results

Before emphasizing the present method capabilities, each part of the optimization process is validated by comparison with reference studies.

39.4.1 Validation Results

Considering a reference test case ($12°$ of incidence, $Re = 10^5$, NACA0012 profile), the comparison of the lift and drag coefficients provided by `Prostar` with previous experimental and numerical values [5] underlines the efficiency of the present method. With $\Delta t = 10^{-2}$ s as time-step, the whole computation lasts approximately 40 min on a XEON 2.4 GHz mono-processor/RAM 3 GB and the evaluation of the complete efficiency graphe lasts 7 h. The comparison between NACA0012 power graphe computed by the present method and experimental results [6] shows a real agreement for the same "solidity" $\left(\frac{\text{number of blades} \times \text{ blade chord}}{\text{rotor radius}}\right)$ value (0.2), considering the static approximation.

39.4.2 Optimization Results

The optimization presented is based on the two parameter shape design space. The evaluated profiles associated to each experiment are illustrated on Fig. 39.1. The objective of the present optimization process is to maximize the nominal power production and the range efficiency under an inequality constraint on blade weight. An important fact is that the RSM related to nominal production presents a local optimum. As a consequence the best way to solve is the Genetic Algorithms approach, which converge in nine iterations.

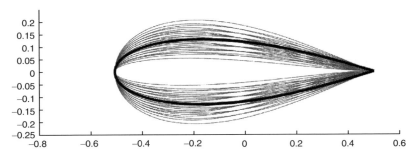

Fig. 39.1. DOE on a two dimension design space and optimal blade profile

The optimal profile is thicker than classical ones (cf. Fig. 39.1) and the gains compared to NACA0015 profile are as follows:

+28%nominal power production($C_\mathrm{p} = 0.41$), +46% efficiency range.

39.5 Conclusion and Prospects

The optimization process described in the present study is based on a parametric DOE/RSM approach which implies a specific interfacing between Optimization and CFD tools. The physical efficiency of the model used can confirm a certain realism of the optimal shape design which is very close to NACA0025 profile.

An improvement of the numerical simulation by a phase-average eddy simulation will accurate the physical model and thus the reliability of the optimization. Furthermore, this improved CFD method will be used to take into account of unsteady effects like dynamic stall and aerodynamical interaction between blades. The next step of this study will consist in implementing other shape design technics like topological analysis and level-set method.

Acknowledgements to NOESIS/LMS for collaboration and EDF R&D, "réseau RNTL"/METISSE and ADEME for financial support.

References

1. I. Paraschivoiu (2002): Wind Turbine Design, with emphass on Darrieus Concept, Polytechnic International
2. R.J. Templin (1974): Aerodynamic Performance Theory for the NRC Vertical Axis Wind Turbine, N.A.E. report, LTR-LA-160
3. Y.S. Chen and S.W. Kim (1987): Computation of turbulent flows using an extended k–ε closure model, NASA CR-179204

4. J. Stoer (1985): Foundation of recursive quadratic programming methods for solving nonlinear programs, Computational Mathematical Programming, NATO ASI Series F: Computer and Systems Sciences, Vol. 15, Springer, Berlin Heidelberg New York
5. L. Smaguina-Laval (1998): Analyse physique et modélisation d'écoulements instationnaires turbulents autour de structures portantes en aérodynamique externe, thèse de l'Université de la Méditerranée – Aix-Marseilles II
6. B.F. Blackwell, R.E. Sheldahl, L.V. Feltz (1976): Wind tunnel performance datas for the Darrieus wind turbine with NACA0012 blades, Sandia Laboratories Report SAND76-0130

40

Rotation and Turbulence Effects on a HAWT Blade Airfoil Aerodynamics

Christophe Sicot, Philippe Devinant Stephane Loyer and Jacques Hureau

40.1 Introduction

Incident flows on wind turbines are often highly turbulent because they operate in the atmospheric boundary layer (ABL) and sometimes in the wake of other wind turbines. In the ABL, the turbulence level varies from 5% (offshore) to 40% and length scale varies typically from 0.001 to 500 m or more [1]. The blade aerodynamics is greatly influenced by rotation [2]. The flow on wind turbine blade being often separated due to the angle of attack encountered which depends on turbulence level, tower shadow or yaw misalignement. The present study is focused on the coupled influence of turbulence and rotation on the separation on the blade airfoil. After presenting the experimental setup, the separation on the blade is studied from average pressure distributions. The high turbulence levels lead to an important measurement dispersion. This is the reason why, in a second part, the instantaneous pressure distributions are analysed.

40.2 Experiment

The wind turbine, tested in the laboratory wind tunnel, was positioned in a 4 m x 4 m section, 1 m downwind of a square 3 m x 3 m nozzle generating a free open jet (Fig. 40.1). An isotropic and homogeneous turbulence, with integral length scale of the order of magnitude of the chord length and three different streamwise intensities (4.4%, 9% and 12%) was generated using a square grid mounted in the wind tunnel nozzle (Fig. 40.1). The wind turbine rotor diameter is 1.34 m, the blades have no twist, a NACA 65_4-421 airfoil, and a 71 mm chord length. Surface pressure data have been obtained with one blade section equipped with 26 pressure taps at three radial positions from the hub : $\frac{r}{R} = 0.26$, 0.51 and 0.75. Pressure were measured simultaneously on the 26 taps at a frequency of 200 Hz during 10.25 s.

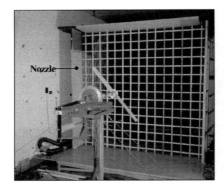 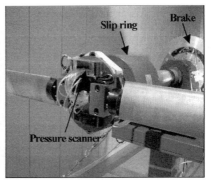

Fig. 40.1. Experimental setup. Note turbulence grid (9%) in nozzle

The angle of attack at a given section of the blade is $\alpha_r = \alpha_{geo} - \alpha_i$ where α_{geo} is a geometrical angle and α_i the angle induced by the wind turbine wake. Whereas α_{geo} can be easily measured, α_i cannot be measured and has been obtained using a lifting line code developed at the laboratory.

40.3 Results and Discussion

40.3.1 Mean Pressure Values Analysis

According to Corten [3], pressure in the separated area is governed, in the chordwise direction by: $\frac{\partial p}{\rho r \partial \theta} = 2 v_r \Omega$ where Ω is the wind turbine rotational velocity. We assumed that the radial velocity in the separated area, v_r, is constant which leads to a constant chordwise pressure gradient. This assumption has been experimentally validated. Thus, the separated area is defined as the part of the average pressure distribution curve where the slope of the pressure gradient is constant (Fig. 40.2). T_{ur} and α_r are, respectively, the local streamwise turbulence level and the local angle of attack on the blade. The pressure gradient and the separation point position, $D_{\overline{Cp}}$, have been obtained. It has also been observed a stall delay for the greatest turbulence level (Fig. 40.3) as it was observed during tests performed at the laboratory on a non-rotating airfoil [4].

The pressure gradient values measured have been compared with semi-empirical values obtained for a non-turbulent flow $\frac{\partial Cp}{r.\partial \Theta} = \frac{4.\sqrt{2}.\lambda^2}{r.(\lambda^2 + (\frac{R}{r})^2)}$ ([5] on the basis of [3]) (Fig. 40.4). Although semi-empirical results are overestimated, probably because of approximation made on radial velocity, they show the same tendency as the experimental ones. The pressure gradient increases with tip speed ratio and turbulence level.

Surface pressure distributions on a rotating and a non-rotating airfoil have been compared (Fig. 40.5). The pressure distribution on the rotating and non-rotating airfoil are similar on the pressure side (attached flow), but a

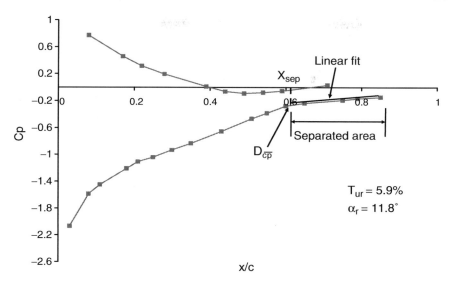

Fig. 40.2. Pressure gradient in separated area

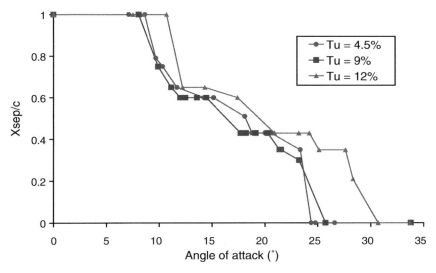

Fig. 40.3. Turbulence level influence on the separation point position, X_{sep}

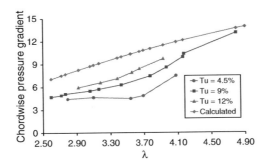

Fig. 40.4. Turbulence level influence on the pressure gradient in the separated area

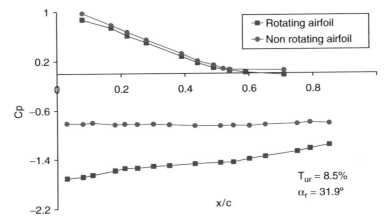

Fig. 40.5. Comparison of surface pressure distribution on a non-rotating and a rotating airfoil

lower pressure on the rotating airfoil suction surface leads to a greater lift, already observed in [2].

40.3.2 Instantaneous Pressure Distributions Analysis

The increase observed in the dispersion of pressure measurements with turbulence level can be quantified analysing the standard deviation of Cp, σ, on the airfoil suction surface (Fig. 40.6).

Separation spreads over a streamwise region [6]. The curve $\sigma = f(\frac{x}{c})$, which can be separated in four main areas, has been analysed in order to quantify the displacement zone of the separation point (Fig. 40.6). The zone 1 shows a strong increase of σ due to the fluctuation of the stagnation point near the leading edge. Zones 2 and 4 show a relatively constant σ and zone 3 shows a bump of σ.

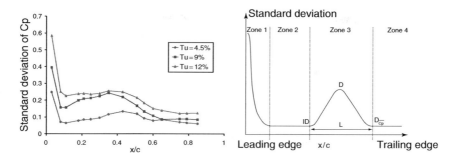

Fig. 40.6. Turbulence level effect on standard deviation and definition of displacement zone from standard deviation curve

The separation point, D, is classically defined as the point where instantaneous backfill occurs 50% of the time [6]. It may then be assumed to be the location where the dispersion, σ, is maximum and $D_{\overline{Cp}}$ may be considered as the location where the flow is always separated which corresponds to the end of the zone 3. The point incipent detachment (ID), at which instantaneous backflow occurs 1% of the time [6], then corresponds to the beginning of the zone 3. Thus, the standard deviation bump can be interpreted, in a first approach, as a representation of the separation point displacement zone and the maximum standard deviation as the middle of this zone.

40.4 Conclusion

Coupled effects of turbulence and rotation on wind turbine airfoil aerodynamics have been studied. A stall delay has been observed for the highest turbulence level. The pressure gradient in the separated area shows an increase with the turbulence level. The instantaneous pressure distributions have been analysed. It has been assumed that the length of the σ pressure coefficient bump may be assimilated with the length of the separation point displacement zone. The method proposed to detect the separation point has to be validated with, for example, results issue from PIV or LDV experiments.

References

1. Kaimal J.C., Finnigan J.J., (1994) Atmospheric Boundary Layer Flows – Their Structure and Measurement. Oxford University Press
2. Schreck S. (2004) Rotational Augmentation and Dynamic Stall in Wind Turbine Aerodynamics Experiments. World Renewable Energy Congress VIII, Denver.
3. Corten G.P. (2001) Flow Separation on Wind Turbine Blades. Thesis, Utrecht University, Utrecht

226   C. Sicot et al.

4. Devinant P., Laverne T., Hureau J. (2002) Experimental study of wind-turbine airfoil aerodynamics in high turbulence. J. Wind Eng. Ind. Aero. 90:689–707
5. Sicot C. (2005) Etude en soufflerie d'une éolienne à axe horizontal. Influence de la turbulence sur l'aérodynamique de ses profils constitutifs. Thesis, Université d'Orléans, Orléans
6. Simpson R.L., Chew Y.T., Shivaprasad B.G (1981) The structure of a separating turbulent boundary layer: part I, mean flow and Reynolds stresses; part II, higher order turbulence results. J. Fluid Mech. 113:23–77

41

3D Numerical Simulation and Evaluation of the Air Flow Through Wind Turbine Rotors with Focus on the Hub Area

J. Rauch, T. Krämer, B. Heinzelmann, J. Twele and P.U. Thamsen

Summary. Blade element and momentum methods (BEM) are the traditional design approach to calculate drag and lift forces of wind turbine (WT) rotor blades. The major disadvantage of these theories is that the airflow is reduced to axial and circumferential flow components [1]. Disregarding radial flow components leads to underestimation of lift and thrust [2]. Therefore correction models for rotational effects are often used. In some situations these models do not describe the strong increase of lift and drag at the hub [3] with sufficient accuracy, so further investigation is necessary.

A 3D numerical simulation is accomplished to map and analyze the flow field of the WT with a focus on the hub area and regarding radial flow components. The knowledge extracted from visualizations of rotational effects on the flow field along the blade can be used to develop a tool for the design optimization of WT rotor blades. In this study the simulation grid is generated with Ansys ICEM 5.1 and the numerical solution is achieved in a block-structured grid consisting of 3×10^6 hexahedron using the commercial computational fluid dynamics (CFD) software CFX 5.7.1. The rotor blade manufacturer EUROS provided the geometric data and the CFD Service Provider CFX-Berlin provided the software support.

Key words: CFD, Rotor Aerodynamics, Rotational Effects

41.1 Introduction

Wind turbine Root profiles of cylindric shape lead to flow separation even at small Reynolds numbers. At this innermost rotor section power production is comparably low but not negligible [4]. Due to massive flow separation radial effects occur and influence the airflow at larger radii.

This study has a focus on the airflow in the hub area and aims to describe its influence on the power production at the example of an Euros "Eu 56" blade of 28.4 m radius operating at rated wind speed of $10.8\,\mathrm{m\,s^{-1}}$.

41.2 Method

The unity of rotor and nacelle was designed as shown in Fig. 41.1. Since effects from the tower are not investigated in this phase of the study it was not implemented within the geometry. As a consequence the 120° symmetry of the three bladed rotor and nacelle could be used. This is a common attempt in turbo machinery simulation and reduces the calculating effort significantly. Hub, nacelle and blade design are based on a Winwind WWD-1 turbine.

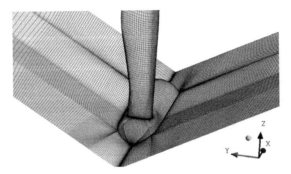

Fig. 41.1. Section of the volume mesh

The upper surface of the domain is designed as an opening boundary condition to allow the typical expansion of the stream tube around the WT.

Within the flow regime the rotor has a 100 m distance upwind and 450 m downwind till the boundaries of the domain, its radius is 100 m. A higher mesh quality was achieved by using three general grid interfaces (GGI) that permit a high grid resolution in the hub area with an reasonable amount of grid elements. The numerical simulation was performed using the Reynolds Averaged Navier–Stokes Equation (RANS) with the standard k–ε model.

41.3 Results

On this basis the aerodynamic power was derived from the pressure field on the rotor blade surface and agrees well with technical specification and measurements. The resulting flow separation area along the blade shows good agreement with observations of the manufacturer. Since k–ε models tend to underestimate flow separation [3], the calculated separated area is smaller than found in reality, which is shown in Fig. 41.4. Through this simulation several effects were observed. Here two of them are described.

The two velocity distributions in Fig. 41.3 both rely on the same velocity scale from 0 to $30\,\mathrm{m\,s^{-1}}$. The domination of radial flow components in the

41 3D-CFD-Simulation of a Wind Turbine Rotor Focusing the Hub Area 229

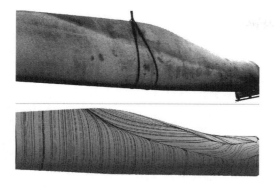

Fig. 41.2. Upper figure: Photography of flow separation effects on EU 56 rotor blade, Lower figure: 3-D Simulation: Surface streamlines on suction side

separated flow can be pointed out by comparing the radial velocity field on the left side to the absolute velocity field on the right side. This is representative for all near-hub-profiles that are affected by flow separation.

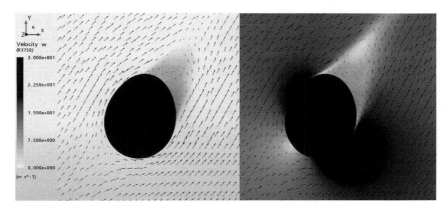

Fig. 41.3. 3D Simulation, Velocity and Vector field for $r = 3.75$ m, left figure: Radial velocity component, right figure: Velocity field

The evaluation of the airflow at different profiles along the rotor blade show an expanding and intensifying radial flow with growing rotor-radii. Beginning at the hub, radial velocities accelerate and remain at $13\,\mathrm{m\,s^{-1}}$ until the rotor ratio of $r/R = 28\%$, where they start to decelerate. At $r/R = 30\%$ all radial flows have been redirected to the trailing edge (see Fig. 41.4).

In Fig. 41.2 the air flow separates at cylindric root profiles characteristically. Two eddies in lee develop differently with growing radius. The analysis of this velocity distribution and those for larger radii indicate a helical movement of the vortex centers.

230   J. Rauch et al.

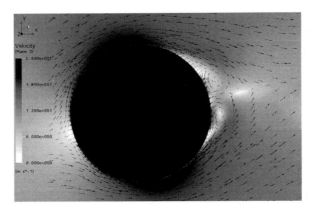

Fig. 41.4. 3D Simulation: Velocity and vector field for radius of 1.5 m

41.4 Perspective

An evaluation of high-resolution 2D-simulations for different profiles is going to be performed. Then a model for comparing 2D and 3D angle of attack has to be chosen. At the moment we prefer the stagnation point method. Afterwards 2D and 3D lift- and drag-coefficients will be calculated and differences will be discussed. Furthermore the vortex system near the hub area is going to be investigated.

References

1. H. Glauert, Airplane Propellers, Vol. 4, Div. L in Aerodynamic Theory, edited by Durand W.F., Dover ed. 1943
2. H. Himmelskamp, (PhD dissertation, Göttingen, 1945); Profile investigations on a rotating airscrew. MAP Volkenrode, Reports and Translation No. 832, September 1947
3. C. Lindenburg, Modelling of rotational augmentation based on engineering considerations and measurements, Energy Research Centre of the Netherlands, ECN-RX–04-131
4. E. Hau, Windkraftanlagen, Springer, Berlin Heidelberg New York, 2002

42

Performance of the Risø-B1 Airfoil Family for Wind Turbines

Christian Bak, Mac Gaunaa and Ioannis Antoniou

Summary. This paper presents the measurements on the Risø-B1 airfoil family with relative thicknesses of 15%, 18%, 24% and 27% tested in the *Velux* wind tunnel. The measurements carried out at a Reynolds number of 1.6×10^6 showed that the airfoils have a high maximum lift and that they are very insensitive to leading edge roughness. This makes the airfoils suitable for design of less solid rotors and still obtain reliable energy production.

42.1 Introduction

The Risø-B1 airfoil family was developed with a relative thickness range from 15% to 53% and with high maximum lift [1]. They were designed for megawatt size wind turbines with pitch control and variable rotor speed (PRVS). For PRVS wind turbines a high maximum lift ($c_{l,max}$) with a corresponding high design lift means that less solid rotors can be obtained. The reduced solidity potentially ensures savings in manufacturing the blades and reduced loads on the remaining components of the turbine. Furthermore, the airfoils were designed for high driving force in the chordwise direction, to be insensitive to leading edge roughness and to have a smooth trailing edge stall despite of the high lift. The design Reynolds number was $Re = 6 \times 10^6$ and was based on a design tool for multi-point airfoil design [2] coupled to *XFOIL*. For evaluation the CFD solver, *EllipSys2D*, [3,4] was used. The design of the series and measurements of the 18% and 24% airfoil were reported in [1] and since then measurements on the 15% and 27% airfoil have been carried out. Therefore, this paper presents the verification of the performance of the 15%, 18%, 24% and 27% Risø-B1 airfoils.

42.2 The Wind Tunnel

The Risø-B1 airfoils were tested in the *Velux* wind tunnel in Denmark. This tunnel and the measuring techniques are described and verified in detail in [5].

232   C. Bak et al.

The measured airfoil configurations presented in this paper is (1) Clean surface with no aerodynamic devices mounted on the airfoil (*Clean*), (2) Standard leading edge roughness (LER) simulated by zigzag tape mounted at the suction side at $x/c = 0.05$ and at the pressure side at $x/c = 0.10$ (*Standard LER*) and (3) Severe LER simulated by zigzag tape mounted at the suction side from the very leading edge towards the trailing edge (*Severe LER*). A description of the test is found in [6].

42.3 Results

The measured characteristics for the Risø-B1 airfoils at $Re = 1.6 \times 10^6$ with clean surface and simulated leading edge roughness are reflected in Fig. 42.1 and Fig. 42.2. Also, corresponding 2D CFD computations at $Re = 1.6 \times 10^6$ are shown. Key parameters reflecting the performance of the airfoils are seen in Table 42.1. Values of maximum lift for the *clean* configuration show that the design objective for high maximum lift was met. The performance with *standard LER*, which probably is the way of simulation of LER that

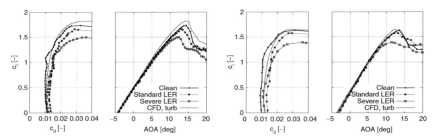

Fig. 42.1. Measurements with clean surface, standard LER and severe LER. The two left plots: The performance for the Risø-B1-15. The two right plots: The performance for the Risø-B1-18. Also, results from 2D CFD computations are seen

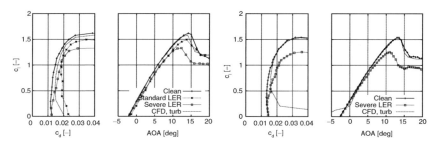

Fig. 42.2. Measurements with clean surface, standard LER and severe LER. Two left plots: The performance for the Risø-B1-24. The two right plots: The performance for the Risø-B1-27. Also, results from 2D CFD computations are seen

42 Performance of the Risø-B1 Airfoil Family   233

Table 42.1. Key parameters for the Risø-B1 family measured at $Re = 1.6 \times 10^6$

Airfoil	Configuration	$c_{l,max}$ (at AOA)	$c_{d,min}$ (at AOA)	max c_l − c_d-ratio (at c_l)
	Clean	1.74 (14.4)	0.0099 (1.4)	106 (1.08)
Risø-B1-15	Standard LER	1.66 (13.0)	0.0125 (−1.7)	74 (1.12)
	Severe LER	1.49 (12.4)	0.0108 (−2.4)	72 (1.10)
	Clean	1.64 (13.5)	0.0090 (−0.6)	100 (1.16)
Risø-B1-18	Standard LER	1.58 (13.2)	0.0130 (−0.2)	71 (1.19)
	Severe LER	1.39 (11.4)	0.0108 (−0.7)	62 (1.02)
	Clean	1.62 (14.2)	0.0117 (−1.0)	73 (1.13)
Risø-B1-24	Standard LER	1.50 (13.7)	0.0184 (5.0)	60 (1.19)
	Severe LER	1.33 (12.3)	0.0121 (−0.8)	58 (1.09)
Risø-B1-27	Clean	1.54 (13.5)	0.0131 (0.5)	64 (1.16)
	Severe LER	1.26 (11.0)	0.0136 (−0.6)	49 (1.05)

Table 42.2. Key parameters for the Risø-B1 family computed using 2D CFD at $Re = 6 \times 10^6$

Airfoil	Configuration	$c_{l,max}$ (at AOA)	$c_{d,min}$ (at AOA)	max c_l − c_d-ratio (at c_l)
Risø-B1-15	Clean	2.00 (16)	0.0088 (−2)	110 (1.40)
Risø-B1-18	Clean	1.90 (15)	0.0090 (−2)	106 (1.37)
Risø-B1-24	Clean	1.82 (16)	0.0112 (2)	91 (1.38)
Risø-B1-27	Clean	1.73 (15)	0.0124 (2)	84 (1.35)

corresponds best to a realistic leading edge roughness on wind turbine rotors, show that the airfoils are very insensitive to leading edge roughness, thus meeting another important design objective. The lift coefficients at which the lift-drag ratio has its maximum, also called the design lift, are relatively high. This makes the airfoils suitable for design of slender blades.

Comparing the measurements in the clean configuration to the 2D CFD computations, show excellent agreement for the 18%, 24% and 27% airfoils and good agreement for the 15% airfoil lift. Estimating the performance of the Risø-B1 airfoil family at $Re = 6 \times 10^6$ using fully turbulent 2D CFD is believed to result in reliable maximum lift values, whereas the drag values in general will be over predicted. Such values are seen in Table 42.2.

42.4 Conclusions

The Risø-B1 airfoil family was tested in the *Velux* wind tunnel. The measurements showed a very good agreement with the predicted performance. The airfoils showed as desired, high maximum lift and low roughness sensitivity.

234 C. Bak et al.

Results from 2D CFD computations at $Re = 6 \times 10^6$, corresponding to megawatt size wind turbines, showed a further increase in maximum lift and in design lift.

The fact that the Risø-B1 airfoils had very high maximum lift and were only slightly sensitive to leading edge roughness can be used for increasing the design lift and thereby designing less solid rotors and still obtain reliable energy production.

42.5 Acknowledgements

The Danish Energy Authority funded this work (ENS 1363/00-0007, "Program for research in aeroelasticity 2000–2001" and ENS 33030-0005, "Program for applied research in aeroelasticity 2004," ENS 33030-0005).

References

1. Fuglsang P, Bak C, Gaunaa M, Antoniou I, Design and verification of the Risø-B1 airfoil family for wind turbines, Solar Energy Engineering **126**, 1002–1010 2004
2. Fuglsang P, Dahl KS, Multipoint optimization of thick high lift airfoil for wind turbines, Proc. EWEC'97, 6–9 October 1997, Dublin, Ireland, pp. 468–471
3. Sørensen NN, General Purpose Flow Solver Applied to Flow over Hills, Risø-R-827(EN), Risø National Laboratory, Denmark, 1995
4. Michelsen JA, Basis3D – a Platform for Development of Multiblock PDE Solvers, Technical Report AFM 92-05, Technical University of Denmark, 1992
5. Fuglsang P, Antoniou I, Sørensen NN, Madsen HA, Validation of a Wind Tunnel Testing Facility for Blade Pressure Measurements, Risø-R-981(EN), Risø National Laboratory, Roskilde, Denmark, 1998
6. Fuglsang P, Bak C, Gaunaa M, Antoniou I, Wind Tunnel Tests of Risø-B1-18 and Risø-B1-24, Risø-R-1375(EN), Risø National Laboratory, Roskilde, Denmark, 2003

43

Aerodynamic Behaviour of a New Type of Slow-Running VAWT

J.-L. Menet

43.1 Introduction

Wind turbines are dimensioned for a nominal running point, i.e. for a given wind velocity. In most cases, principally because of their higher efficiency, two- or three-bladed fast running wind turbines are preferred [1].

Although it has undeniably high performances, this type of wind-powered engine is not necessarily the one which enables the extraction of the maximal energy from a wind site. Using a special method, namely the L–σ criterion, Menet et al. [2] have shown that a slow-running vertical axis wind turbine (VAWT), such as the Savonius rotor [3], can extract more energy than fast running wind machines.

This idea seems to be in contradiction to the general literature in the field: the Savonius rotors have an aerodynamic behaviour where the characteristics of a drag device dominate, which clearly induces a low efficiency. In fact, using the same intercepted front width of wind **L** and the same value σ of the maximal mechanical stress on the paddles or the blades, the reference [2] clearly indicates that the delivered power of a Savonius rotor is superior to the one of any fast-running horizontal axis wind turbine. Besides, because of its high starting torque, a Savonius rotor can theoretically produce energy at low wind velocities, and because of its low angular velocity, it can deliver electricity under high wind velocities, when fast running wind turbines must generally be stopped [1, 4].

The main disadvantage of the Savonius rotor is the great instability of the mechanical torque because the flow inside the rotor is non-stationary. Nevertheless, the advantages of such a wind turbine are numerous [4]. For a few years, many studies have led to raise performances of these machines [5–7]. Here is the purpose of our contribution to the study.

Fig. 43.1. 3D representation

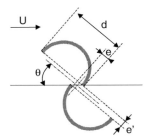

Fig. 43.2. Savonius rotor

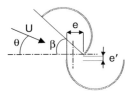

Fig. 43.3. Modified rotor

43.2 Description of the Savonius Rotors

The Savonius rotor is made from two vertical half-cylinders running around a vertical axis (Figs. 43.1, 43.2).

A *modified rotor* (Fig. 43.3) is also studied, which is just an extrapolation of the Savonius rotor, using three geometrical parameters: the main overlap e, the secondary overlap e', and the angle β between the paddles.

In the following, we name *conventional Savonius rotor* the one for which the geometrical parameters e and e' are, respectively, equal to $d/6$ and 0. The characteristic curves of such a rotor (values of the power coefficient C_p and of the torque coefficient C_m towards the speed ratio λ) are presented in Fig. 43.4.

43.3 Description of the Numerical Model

The model used in this study is a mono-stepped rotor of infinite height. This hypothesis has allowed to carry out the calculations on the two-dimensional

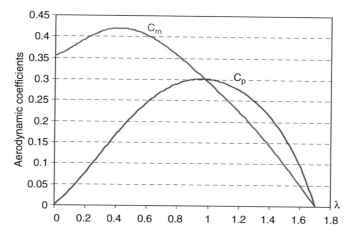

Fig. 43.4. Aerodynamic performances of the *conventional Savonius rotor* [8]

model represented in Fig. 43.2. The three geometric parameters e, e' and θ are selected for the numerical study.

The influence of the dynamic parameter, namely the Reynolds number Re_D, based on the rotor diameter D of the rotor, has also been investigated [7]. Only the influence of the overlap is presented in this paper.

The finite volume mesh obtained is about 50,000 cells large and made of quadrilaterals. First-order discretisation schemes have been used for the pressure and the velocity computations of an incompressible flow, together with a high Reynolds number two equations turbulence model.

The calculations have been realised using a static calculation (rotor supposed to be fixed whatever the wind direction θ) and also a dynamic calculation. Only the static calculation is presented here.

43.4 Results

For the present calculation, the Reynolds number corresponds to the nominal conditions for the prototype [6]: $Re_D = 1.56 \times 10^5$.

43.4.1 Optimised Savonius Rotor

A static simulation of the flow around few Savonius rotors, allowed to determine the pressure distribution on the paddles for different values of the wind direction θ. (c.f. example on Fig. 43.5). Then the static torque is calculated as a function of θ (Fig. 43.6).

The present simulation gives satisfactory results since the differences between the numerical simulation and the experimental data [9] are inferior to 10% for most angles θ. The optimal value for the relative overlap ratio e/d

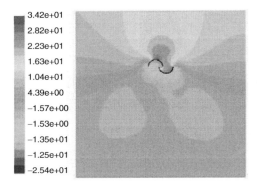

Fig. 43.5. Pressure contours (Pa) on the *optimised rotor* ($\theta = 90°$)

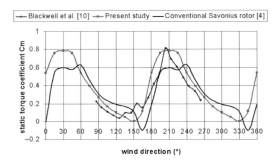

Fig. 43.6. Static torque coefficient on the *optimised rotor*

is found to be equal to 0.242, whereas $e' = 0$. Even if it remains very instable, the torque coefficient is notably raised compared with the *conventional Savonius rotor* [4], with a general increasing of about 20%.

43.4.2 The New Rotor

In this preliminary work, the influence of the inclination angle β of the paddles has been studied for only two values of the wind direction: $\theta = 90°$ and $\theta = 45°$, that allows to determine an optimal value of the inclination angle: $\beta = 55°$. Optimum values of the overlaps giving highest values of the torque, have been systematically researched, leading to: $e/d = 0.242$ and $e' = 0$. The static torque is calculated as in Sect. 43.4.1. (Fig. 43.7). An example of the pressure contours is presented in Fig. 43.8.

The result is encouraging, since the *new rotor* induces maximal values of the static torque largely higher than those obtained on the *conventional rotor*, even if it also introduces low and negative values of this torque, with a great angular variation. Nevertheless, the mean value of the torque is increased: $C_m = 0.48$, i.e. 60% more than for the *conventional rotor*.

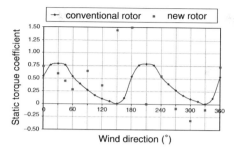

Fig. 43.7. Torque on the *optimised rotor* ($\theta = 90°$; $\beta = 45°$; $e/d = 0.242$; $e' = 0$)

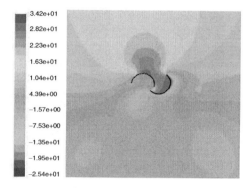

Fig. 43.8. Pressure contours (Pa) on the *optimised rotor* ($\theta = 45°$)

The choice of the three geometrical parameters e, e' and θ should be made simultaneously not only to increase the aerodynamic efficiency but also to aim an angular stability of the torque (future studies).

43.5 Conclusion

The flow around few *conventional Savonius rotors* has been modellised using the CFD code Fluent v6.0. A geometry of an *optimised rotor* has been proposed. Solutions have been presented to invent a *new rotor*, with higher performances, by acting on the overlaps of the paddles and on their relative inclination. This work must be continued, aiming a simultaneously research of the optimum values of these geometrical parameters.

References

1. Wilson RE, Lissaman PBS. (1974) Applied Aerodynamics of wind power machines. Research Appl. to Nat. Needs, GI-41840, Oregon State University
2. Menet JL, Valdès LC, Ménart B (2001) A comparative calculation of the wind turbines capacities on the basis of the L–σ criterion. Ren. Eng. 22: 491–506

3. Savonius SJ (1931) The S-rotor and its applications. Mech. Eng. 53-5: 333–337
4. Martin J (1997) Energies éoliennes (in French). Techniques de l'Ingénieur B1360, France
5. Ushiyama I, Nagai H (1988) Optimum design configurations and performances of Savonius rotors. Wind Eng. 12-1: 59–75
6. Menet JL (2004) A double-step Savonius rotor for local production of electricity: a design study. Ren. Eng. 29: 1843–1862
7. Menet JL, Bourabaa N (2004) Increase in the Savonius rotors efficiency via a parametric investigation. In : European Wind Energy Conference, London
8. Le Gourières D (1980) Énergie éolienne, Eyrolles, Paris : chapter V
9. Blackwell BF, Shedahl RE, Feltz LV (1977) Wind tunnel Performance data for two- and three-bucket Savonius rotors. Final report SAND76-0131, Sandia Lab., CA

44

Numerical Simulation of Dynamic Stall using Spectral/hp Method

B. Stoevesandt, J. Peinke, A. Shishkin and C. Wagner

Reduction of the weight of wind turbine blades is of large interest to manufacturers. To do so without losing the necessary stability, blade loads caused by dynamic stall effects have been a major task in aerodynamic research during the last years [1].

The aim of the project is to calculate lift and drag caused by the effect of dynamic stall as they arise in turbulent wind fields. This is done by means of wind tunnel measurements and numerical flow simulations. For the flow simulation a spectral/hp code has been chosen to achieve high accuracy [2]. In order to simulate the dynamic stall the boundary conditions have to be flexible. Currently the main focus of the works lies on reliability tests of the numerical code.

44.1 Introduction

Due to the increasing size of wind turbine airfoils their weight is becoming an increasing problem. The aim of highest efficiency, best lift at a minimum of cost and weight seems to be an unsolvable contradiction. One of the problems in the design of airfoils is, that mechanical stability investigations are based on estimations of the lift caused by dynamic stall.

Dynamic stall is induced by an unsteady inflow on the airfoil. For wind turbines a main factor is gusty inflow on the profile causing rapid changes of wind speed and direction. Thus at a tip with a tip speed of $80\,\mathrm{m\,s^{-1}}$ a change of wind speed of $8\,\mathrm{m\,s^{-1}}$ would even cause an inflow deviance of $5.7°$. Figure 44.1 illustrates that these occurrences are of relevant order.

Simulating turbulence is difficult with traditional CFD approaches since these methods are stabilized using numerical dissipation. In contrast the numerical dissipation of spectral/hp-element methods is comparably low. The aim of the project is to investigate the flow separation at the profile in 3D and to analyse the extreme loads on the airfoil caused by turbulence.

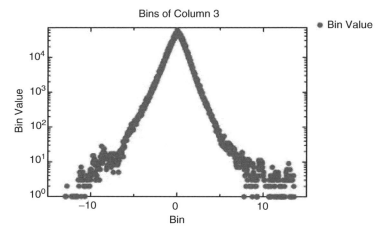

Fig. 44.1. PDF of Windspeed change within 2 s within a 22 h measurement at Tjare (by Böttcher, data from http://www.winddata.com)

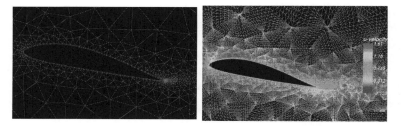

Fig. 44.2. Typical mesh on a fx79-w-151 profile (*left*) and nodal points generated (*right*)

44.2 The Spectral/hp Method

The spectral/hp element method combines the accuracy of classical spectral codes with the flexibility of finite element methods (FEM). Global spectral methods use one representation of a function $u(x)$ by a series through the complete domain. This idea is advanced by the spectral/hp method by subdividing the domain into finite – unstructured – elements. This is done by another transformation on the spectral function. To fulfill the necessary requirements, the spectral functions of the separate elements have to be C^0 continuous.

The subdivision into a mesh in combination with the use of unstructured meshes allows an adaption of the problem to the curved geometry and a well tuned resolution to the flowfield.

There are two ways to achieve convergence using the spectral/hp method. The h-type convergence refers to convergence due to a refinement of the mesh, the p-type expansion refers to the order of the polynomial expansion of the function and to the convergence by selecting a sufficient polynomial order. The polynomials can be of nodal type such as Lagrange polynomials or modal type like Jacobi polynomials.

44.3 The NekTar Code

For the numerical simulation we use the NekTar code,[1] which was originally developed by Sherwin, Karniadakis, Warburton et al. It is a scientific code consisting of a 2D and a 3D direct numerical simulation (DNS) solver as well as a 3D-Fourier code including an large eddy simulation (LES) solver. The general option for variable boundary conditions implemented enables the simulation of turbulence .

Figure 44.3 shows a calculation of the flow around an airfoil in 2D.

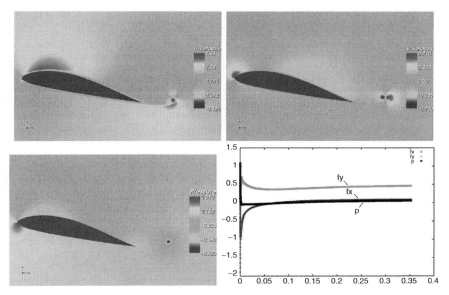

Fig. 44.3. Example of horizontal and vertical flow velocity, pressure and force simulated on a fx79-w-151-airfoil

---

[1] The DNS-NekTar-Codes can be obtained from:
http://www.ae.ic.ac.uk/staff/sherwin/nektar/downloads/

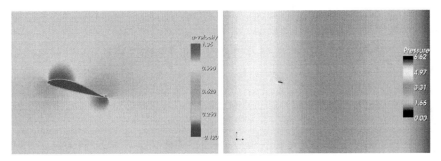

Fig. 44.4. u-velocity at time variable boundary conditions (*left*) and suspicious pressure output (*right*) throughout the domain

44.4 First Results

The 2D simulation worked well under stationary inflow. As the boundary conditions were set to periodic changes of the inflow boundaries, some difficulties were encountered. Obviously the calculation process in Nektar is working for the velocity flows (see Fig. 44.4 left). Therefore the general process of solving the problems seems to be working well. Nevertheless the pressure output shows relatively high pressure variations (Fig. 44.4 right). As the force on the profile seems to be in the same order as for the stationary calculation, the general calculation is obviously correct. These results are in need of further verification.

44.5 Outlook

In a first step the pressure calculation for the time variable boundary condition has to be improved. Especially for simulating variable inflow it seems to be reasonable to implement a new way to establish time variable boundary conditions into the code. Since so far the time variability is limited to continuous functions, it seems convenient to create a more flexible code for time variable inflow.

As so far only the 2D-DNS code has been tested, all the further extension have to be implemented in the 3D- and Fourier-LES code. As DNS is slow, simulation has been done so far for low Reynolds numbers only.

References

1. Amandolese, X., Szechenyi, E.: Experimental Study of the Effect of Turbulence on a Section Model Blade Oscillating in Stall. Wind Energy 7: 267–282, 2004
2. Karniadakis, G. E. and Sherwin, S. J.: Spectral/HP Element Methods for CFD. Oxford University Press, Oxford, 1999

45

Modeling of the Far Wake behind a Wind Turbine

Jens N. Sørensen and Valery L. Okulov

When wind turbines are clustered in a wind farm the power production will be reduced due to the reduced free-stream velocity. However, an even more important impact on the economics of a wind farm is the increased turbulence intensity from the wakes of the surrounding wind turbines that increases the fatigue loadings. Even though many wake studies have been performed over the last two decades [1], a lot of basic questions still need to be clarified in order to elucidate the structure and dynamic behavior of wind turbine wakes.

In the present work various vortex models for far wakes are analyzed, resulting in the development of a new analytical wake model. In the model the vortex system is replaced by N helical tip vortices of strength (circulation) Γ embedded in an inner vortex structure representing the vortices emanating from the inner part of the blades and the hub. The developed model forms the basis for a supplementary work on stability of the tip vortices in the far wake behind a wind turbine by Okulov and Sørensen [2].

45.1 Extended Joukowski Model

As basic hypothesis for describing the far wake it is assumed that the wake is fully developed and that the vorticity is concentrated in N helical tip vortices and a hub or root vortex lying along the system axis. This $(N+1)$-vortex system, proposed originally by Joukowski in 1912 for a two-bladed propeller [3], consists of an array of infinitely long helical vortex lines with constant pitch and radius (see Fig. 45.1a). The model is the simplest way of describing the far wake behind a propeller or a wind turbine. The axial velocity of the vortex system is defined in helical variables $(r, \chi = \theta \mp z/l)$ by the formula

$$u_z(r, \chi) = V_0 - w_z(r, \chi), \tag{45.1}$$

where the axial and azimuthal velocity components, w_z and w_θ, respectively, induced by the N helical tip vortices, may be determined by the following expressions, derived by Okulov [4]:

$$w_z(r,\chi) \cong \frac{\Gamma N}{2\pi l}\binom{1}{0} + \frac{\Gamma}{2\pi l}\frac{\sqrt[4]{l^2+a^2}}{\sqrt[4]{l^2+r^2}}Re\sum_{n=1}^{N}\left[\frac{\{\pm\}e^{i(\chi-2\pi n/N)}}{e^{\{\mp\}\xi}-e^{i(\chi-2\pi n/N)}}\right.$$
$$\left.+\frac{l}{24}\left(\frac{3r^2-2l^2}{(l^2+r^2)^{3/2}}+\frac{9a^2+2l^2}{(l^2+a^2)^{3/2}}\right)\ln\left(1-e^{\xi+i(\chi-2\pi n/N)}\right)\right] \quad (45.2)$$
$$w_\theta(r,\chi) = (\Gamma N/2\pi r) - (lw_z(r,\chi)/).$$

Here the terms in the braces are defined such that the upper one corresponds to $r < a$ and the lower one to $r > a$. The helical vortex parameters a, l, Γ are introduced in Fig. 45.1, where Γ_T refers to the circulation of a tip vortex and Γ_0 refers to the circulation of the inner vortex, and

$$e^\xi = \frac{\left[r\left(l+\sqrt{l^2+a^2}\right)\exp\left(\sqrt{l^2+r^2}\right)\right]}{\left[a\left(l+\sqrt{l^2+r^2}\right)\exp\left(\sqrt{l^2+a^2}\right)\right]}.$$

From (45.2) it is seen that the azimuthally averaged axial velocity is constant in the wake. This is a direct implication of one of the basic assumptions of Joukowski, that the circulation along each rotor blade is constant ($\equiv \Gamma$) and that the total circulation is zero. If we look at experimental data, however, a constant axial velocity profile is rarely seen [1,5].

In order to develop a model that is capable of reproducing measured velocity distributions, we extend the model of Joukowski by assuming the tip vortices to be embedded in an axisymmetric helical vortex field formed from the circulation Γ_0 of the rotor blades and the hub (see Fig. 45.1b). The resulting axial and azimuthal velocity components are described by the following formulas:

$$u_z = V_0 - w_z + \frac{\Gamma_0 f}{2\pi l}, \quad u_\theta = w_\theta - \frac{\Gamma_0 f}{2\pi r}, \quad f = \frac{1}{\delta^2}\left\{\begin{matrix}\delta^2 - r^2, & r < \delta \\ 0, & r \geq \delta\end{matrix}\right\}. \quad (45.3)$$

In fact, several possibilities for appropriate velocity distributions exist. Basically, they have to be a solution for the Euler equations and at the same

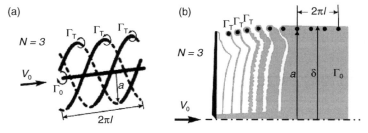

Fig. 45.1. Sketch of the vortices in a far wake behind a rotor: (**a**) model proposed originally by Joukowski (1912); (**b**) vorticity distribution in meridional cross-section of the wake behind a wind turbine rotor

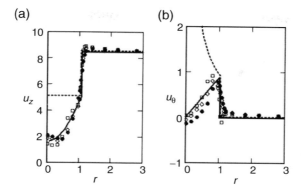

Fig. 45.2. Average axial (**a**) and azimuthal (**b**) velocity profiles calculated by extended model (*boldline*) and by the Joukowski model (*dashedboldline*). The symbols are experimental data measured by Medici & Alfredsson [5] in different cross-sections downstream from the rotor plane: $z/a = 2.8$ (*square*); 3.8 (*diamond*) and 4.8 (*bullet*)

time they need to fit to experimental observations. The proposed model obeys both properties. To validate the extended model (45.3) we compare it to the experimental results of Medici and Alfredsson [5]; see Fig. 45.2. A very good agreement between modeled and measured velocity distributions are seen for both the axial and azimuthal velocity components. In contrast to this, the Joukowski model (45.1) results in an unrealistic $1/r$ behavior for the azimuthal velocity and a constant for the axial velocity.

45.2 Unsteady Behavior

An interesting finding in the measurements of [5] was the appearance of a low frequency in the wake. The frequency was an order of magnitude smaller than that of the rotational frequency of the tip vortices. Medici and Alfredsson attributed the phenomenon to be similar to the vortex shedding occurring in the wake behind a solid disc determined from the Strouhal number. We will utilise their experimental results to validate the developed model. In Fig. 45.3a we depict a meandering wake subject to low-frequency oscillations. The measured time signal of the axial velocity (Fig. 45.3c) near the boundary of the wake is dominated by two frequencies with very high peaks superposed the highest frequency. Such amplifications may be of crucial importance for the lifetime of wind turbines located fully or partly in a wake. By adjusting the amplitude of the low frequency to the measurements we compute a time signal of the velocity (Fig. 45.3d) by de-centering the extended model (45.3) as shown in Fig. 45.3b. The resemblance of the two time histories is striking.

Hence, with the new model, it is possible to model and predict unsteady behavior of far wakes behind wind turbines.

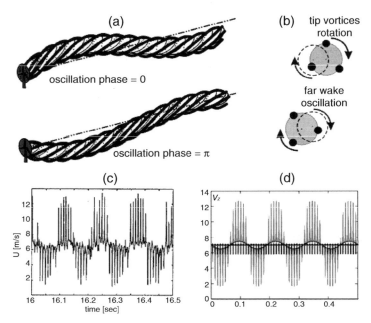

Fig. 45.3. Artistic impression of low-frequency meandering of the wake behind a wind turbine; (**a**) in space and (**b**) in the Trefftz plane. Time histories of the axial velocity; (**c**) measured [5] and (**d**) computed

45.3 Conclusions

The classical vortex model of Joukowski has been extended by specifying an inner vortex structure. The new model enables us to calculate induced velocities in very good agreement with measurements. The model may further explain the very high peaks appearing in the time signal of the axial velocity measured near the center of the tip vortices of unsteady far wakes.

The work was supported by the Danish Energy Agency under the project "Dynamic Wake Modeling" and the RFBR (no. 04-01-00124). The authors thank David Medici for providing the experimental data.

References

1. Vermeer L.J., Sørensen J.N., Crespo A. (2003) Progress in Aerospace Sciences 39: 467–510
2. Okulov V.L., Sørensen, J.N. (2006) Proceedings of the Euromech Symposisum on Wind Energy. This issue
3. Margoulis W. (1922) NACA Tech. Memor. No. 79
4. Okulov V.L. (1995) Russ. J. Eng. Thermophys. 5: 63–75
5. Medici D., Alfredsson P.H. (2004) Proceedings of the Science of Making Torque from Wind, Delft, 2004

46

Stability of the Tip Vortices in the Far Wake behind a Wind Turbine

Valery L. Okulov and Jens N. Sørensen

Summary. The work is a further development of a model developed by one of the authors Okulov (J. Fluid Mech. 521:319–342 2004) in which linear stability of N equally azimuthally spaced infinity long helical vortices with constant pitch were considered. A multiplicity of helical vortices approximates the tip vortices of the far wake behind a wind turbine. The present analysis is extended to include an assigned vorticity field due to root vortices and the hub of the wake. Thus the tip vortices are assumed to be embedded in an axisymmetric helical vortex field formed from the circulation of the inner part of the rotor blades and the hub. As examples of inner vortex fields we consider three generic vorticity distributions (Rankine, Gaussian and Scully vortices) at radial extents ranging from the core radius of a tip vortex to several rotor radii.

46.1 Theory: Analysis of the Stability

The influence of an assigned flow on the stability of multiple vortices has so far only been studied for circular arrays of point vortices or straight vortex filaments (the limiting case of a helical vortex with infinite pitch) [2,3]. A non-perturbed multiple of helical vortices (Fig. 46.1a rotates with constant angular velocity $\Omega = \Omega_0 + \Omega_{\mathrm{Ind}} + \Omega_{\mathrm{Sind}}$ and moves uniformly along its axis with velocity $V = V_0 + V_{\mathrm{Ind}} + V_{\mathrm{Sind}}$. Note that in contrast to the plane case, where the total vortex motion consists of an assigned flow field (Ω_0, V_0) and the mutual induction of the vortices $(\Omega_{\mathrm{Ind}}, V_{\mathrm{Ind}})$ with $V_{\mathrm{Ind}} = 0$, we also include self-induction of each of helical vortex $(\Omega_{\mathrm{Sind}}, V_{\mathrm{Sind}})$ [1]. From linear stability, introducing infinitesimal space displacements $r_k = \alpha(t)\exp(2\pi iks/N)$, $\chi_k = \beta(t)\exp(2\pi iks/N)$, of the k-vortex from the its equilibrium position$(a, 2\pi k/N + \Omega t, Vt)$ a correlation equation $B = 0$ was derived for dimensionless pitch $\tau = l/a$ and vortex radius $\epsilon = r_{\mathrm{core}}/a$

250 V.L. Okulov and J.N. Sørensen

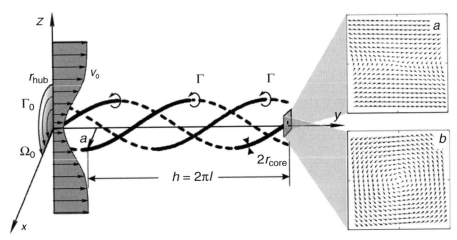

Fig. 46.1. Sketch of N tip helical vortices embedded in assigned flow (Ω_0, V_0) and relative velocity plots about vortex axis for (**a**) unstable and (**b**) stable far wakes

$$\begin{aligned}\frac{4\pi a^2}{\Gamma}B &= s\,(N-s)\,\frac{(1+\tau^2)^{3/2}}{\tau^3} - 2N + 2\frac{N-2}{\tau^2} \\ &+ \frac{1}{\tau(1+\tau^2)^{3/2}}\left[\left(\tau^2 - \frac{1}{4}\right)\left(E + \psi\left(-\frac{s}{N}\right) - \frac{N}{s}\right)\right. \\ &\left.+ \frac{3}{4} - 2\tau^2 - \ln\left(N\varepsilon\frac{(1+\tau^2)^{3/2}}{\tau}\right)\right] + \frac{1+2\tau^2}{\tau(1+\tau^2)^{1/2}} \\ &+ \frac{\tau^3}{(1+\tau^2)^{9/2}}\left(2\tau^4 - 6\tau^2 + \frac{3}{4}\right)\frac{1.20206}{N^2} + \frac{4\pi a^2}{\Gamma}B_A \end{aligned} \quad (46.1)$$

where $s = [0;\,N-1]$ is the subharmonic wave number; $E = 0.5772\ldots$ is the Euler constant and the psi-function $\psi(\cdot)$ takes values such as $\psi(-1/2) = 0.0395\ldots$, $\psi(-1/3) = 1.6818\ldots$, $\psi(-2/3) = -1.6320\ldots$ (see, e.g. [1]). The last term of B_A in (46.1) describes the influence of the assigned velocity field due to root vorticity and the hub of the rotor. Three important examples of the central axisymmetrical helical vortices have been considered with R- (step-shape), L- (Gaussian) and S- (hat-type) axial vorticity distribution corresponding to Rankine, Lamb and Scully vortices at $\tau \to \infty$ [4]. For those types of assigned flow field the B_A-term takes form:

$$B_A = 4N\gamma \begin{cases} -1, & r < \delta, \\ 1/\tau^2\delta^2, & r \geq \delta, \end{cases} \quad (2\text{-R})$$

$$B_A = 4N\gamma \frac{1}{1+\delta^2}\left[\frac{1+\tau^2}{\tau^2}\frac{\delta^2}{1+\delta^2} - 1\right], \quad (2\text{-L})$$

46 Stability of the Tip Vortices in the Far Wake behind a Wind Turbine   251

$$B_A = 4N\gamma \left[\left(\frac{1 + \tau^2}{\delta^2 \tau^2} + 1 \right) \exp\left(-\frac{1}{\delta^2} \right) - 1 \right].$$ (2-S)

where $\gamma = \Gamma_0/N\Gamma$ is the circulation ratio i.e. the central (hub) vortex strength normalized by the total multiple circulation of the tip vortices. The dimensionless hub vortex pitch τ keeps the same value as that of the multiples and the hub vortex radius $\delta = r_{hub}/a$ is normalized by a, the radius of cylinder supporting helical vortex axis of the multiple. Note that B_A does not contain any term with wind speed implying that translatory motion from wind does not influence the stability.

The stability condition of the tip vortex configuration only depends on the sign of B. The vortex system is unstable if $B \geq 0$, where the relative velocity field in the vortex position has an unstable critical point (Fig. 46.1a) and for $B < 0$ the point is stable (Fig. 46.1b).

46.2 Application of the Analysis

Analysis of (46.1) shows that a model in which the vortex system is replaced by N tip helical vortices of strength Γ and a line root vortex ($\delta = 0$) of strength $-N\Gamma$ (Joukowski's model) is unconditional unstable. This, however, seems to in conflict with numerous visualizations [5]. The most likely explanation for this apparent contradiction is that the model of Joukowski is too simplified to describe all types of rotor flows. Indeed Joukowski's model should be considered as special case with a constant circulation distribution along the blade span. Contrary to this the (2-R), (2-L) and (2-S) hub models with a finite core radius are more realistic to describe far wake properties for real operating regimes of wind turbines. Their analysis shows that the stability of tip vortices to a large extent depends on the radial extent of the hub vorticity as well as of the type of vorticity distribution. In order to evaluate the influence of core size for both tip and hub vortices, neutral curves for equilibrium helical vortex arrays are plotted in Fig. 46.2. Note that increasing the hub vortex size leads to loss of stability and that increasing the size of tip vortices makes the vortex system more stable. Increasing the number of vortices in the arrays leads to less stability. Qualitatively this agrees with the results of visual observations. However, more experimental data is needed in order to validate the model more thoroughly.

46.3 Conclusions

In the present work the stability problem of a multiple of N helical vortices to infinitesimal space displacements is generalized to include the hub vorticity distribution. The solution allows us to provide an efficient analysis of some of the experimentally observed stable vortex arrays in far wake behind multiblade wind turbines.

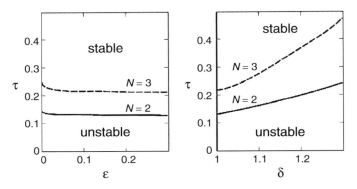

Fig. 46.2. Neutral curves for far-wake with different numbers of tip vortices N as function of helical pitch τ and core size of tip vortices ϵ and hub vortex δ

The work was supported by the Danish Energy Agency under the project "Dynamic Wake Modeling" and the RFBR (no. 04-01-00124).

References

1. Okulov V.L. (2004) J. Fluid Mech. 521: 319–342
2. Havelock T.H. (1931) Philos. Mag. 11: 617–633
3. Morikawa G.K., Swenson, E.V. (1971) Phys. Fluids 14: 1058–1073
4. Kuibin, P.A., Okulov, V.L. (1996) Thermophys. & Aeromech. 3(4): 335–339
5. Vermeer L.J., Sørensen J.N., Crespo A. (2003) Progress in Aerospace Sciences 39: 467–510

47

Modelling Turbulence Intensities Inside Wind Farms

Arne Wessel, Joachim Peinke and Bernhard Lange

Summary. For the turbine design and optimization of wind farm layouts with regard to increased loads the turbulence intensity incident on the wind turbines inside wind farms is needed. Here we present a semi-empirical approach for calculating these turbulence intensities. The approach was compared to horizontal multiple wake turbulence intensity profiles from the Vindeby offshore wind farm.

47.1 Description of the Model

47.1.1 Single Wake Model

The turbulence intensity in the wake of a wind turbine has two origins: the ambient turbulence intensity I_{amb} and the wake induced turbulence intensity I_{add}. The single wake model describes the turbulence intensity profile of the added turbulence intensity I_{add}.

The main effect for the generation of turbulence in the far wake has its origin in the wind speed gradient of the wake. The added turbulence is calculated from two contributions, a wind shear dependent and a diffusion dependent term

$$I_{\mathrm{add}}(r, x) = A I_{\mathrm{mean}}(x) \frac{\partial \tilde{u}(r, x)}{\partial \frac{r}{R}} + B \tilde{u}(r, x) \tag{47.1}$$

Both are based on the normalized wind speed deficit profile $\tilde{u}(r) = 1 - u(r)/u_0$, which describes the wind speed deficit profile at a lateral distance x from the upwind turbine and radial distance r from the center of the wake. For the calculation of the wind speed deficit profile FLaP program [1] with a wake model based on Ainslie [2] is used. u_0 means the incoming free wind speed. $I_{\mathrm{mean}}(x)$ is an approach from Lange [1], deriving a mean turbulence intensity in the wake from the eddy viscosity calculated in the Ainslie model

$$I_{\mathrm{mean}} = \epsilon \frac{2.4}{\kappa u_0 z_{\mathrm{H}}} \tag{47.2}$$

254 A. Wessel et al.

where κ is the von Karman constant (set to 0.4) and z_H the height above the ground.

47.1.2 Superposition of the Wakes

Inside a wind farm, the downwind turbines may be subject to multiple wakes on the rotor from the upwind turbines. The model has to superimpose the wakes from the upwind turbines.

This is done by adding the turbulence intensities of the incident wake I_{add} quadratically and then add the ambient turbulence intensity I_{amb}

$$I = I_{amb} + \sqrt{\sum_{i=1}^{N} I_{add,i}^2}, \qquad (47.3)$$

where N is the number of upwind turbines. The effect of the incoming turbulence intensity on the wake development of a turbine inside a wind farm is taken into account by using the modelled turbulence intensity from the upwind wind turbines as ambient turbulence for the Ainslie model.

47.2 Comparison of the Model with Wake Measurements

The inertial parameters A and B of (47.1) are estimated by a fit of the model to turbulence intensity profile from the Nibe [3] onshore wind farm, described in [4]. The values are $A = 1.42$ and $B = 0.54$.

The model has been compared with measurements of the onshore wind farms Nibe and Sexbierum with good result [4]. Here a comparison to the offshore wind farm Vindeby is performed. Fixed values were used for the parameters A and B for all wind farms with good results. Although in general it could be expected, that they depend on the wind turbine and specific site.

47.2.1 Vindeby Double and Quintuple Wake

The model was compared with the situation at the Vindeby offshore wind farm with double and quintuple wake measurements.

The wind farm is located off the northwestern coast of the island of Lolland, Denmark. The 11 Bonus 450 kW turbines are arranged in two rows as shown in Fig. 47.1. The Bonus 450 have a rotor diameter of 37 m and a hub height of 35 m [5].

Outside the wind farm the two measurement mast SMS and SMW are installed. For a wind direction of 75° a double wake situation occurs for mast SMW while the mast SMS is in the free flow. A quintuple wake is measured at mast SMS at a wind direction of 320°, where SMW captures the free flow conditions.

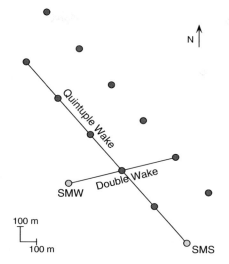

Fig. 47.1. Layout of Vindeby wind farm

The horizontal turbulence intensity profiles were measured by a mast sited in the wake of a wind turbine. Depending on the incoming wind direction, one section of the wake could be measured. For the average periods (normally 10 min), the mean wind directions is used to estimate the measured part of the wake. Bin-averaging is used to get a stationary turbulence intensity profile.

The input wind speed for the measured wind profiles is between 8.5 and $10.5\,\mathrm{m\,s^{-1}}$. The ambient turbulence intensity was measured at the free standing meteorological mast for each direction bin. It is used direction depending for the model (mean value = 5.8%).

As can be seen in Figs. 47.2a,b the modelled data agree well the measured turbulence intensity profiles. Both the maximum turbulence intensity and the profile shape are modelled accurately. The shape of the measured turbulence profiles for double and quintuple wake situation are nearly symmetric, this results from a nearly uniform turbulence intensity over the wind direction due to the homogenous roughness of the surrounding water surface.

47.3 Conclusion

A new model for the turbulence intensity inside wind farms was developed and applied to the Vindeby offshore wind farms.

Horizontal turbulence intensity profiles are compared. The calculated profiles agree very well with the measurements. Both the maximum values and the nearly symmetrical shapes of the measured profiles were modelled accurately.

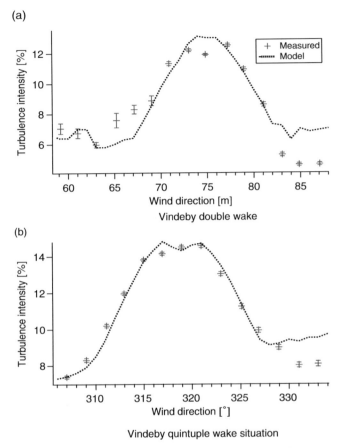

Fig. 47.2. Horizontal turbulence intensity profiles from multiple wake situation at Vindeby wind farm. For the double wake situation a wind direction depending ambient turbulence intensity was used

Acknowledgement

In this work, one of the authors (Arne Wessel) is funded by the scholarship program of the German Federal Environmental Foundation. We thank Risø National Laboratory Denmark, for placing the measurement data from Vindeby offshore wind farm at our disposal.

References

1. B. Lange, H-P. Waldl, A. G. Guerrero, D. Heinemann, and R. J. Barthelmie. Modelling of OffshoreWind turbine wakes with the wind farm program FLaP. *WIND ENERGY*, 6:87–104, 2003

2. J. F. Ainslie. Calculating the flowfield in the wake of wind turbines. *Journal of Wind Engineering and Industrial Aerodynamics*, 27:213–224, 1988
3. G. J. Taylor. Wake measurements on the nibe wind turbines in denmark. Technical report, Risø National Laboratory, 1990
4. A. Wessel and B. Lange. A new approach for calculating the turbulence intensities inside a wind farm. In *Proceedings of the Deutschen Windenergiekonferenz (DEWEK)*, Wilhelmshaven, 2004
5. R. J. Barthelmie, M. S. Courtney, J. Højstrup, and P. Sanderhoff. The Vindeby project: A description. Technical Report Risø-R-741(EN), RISØ, 1994

48

Numerical Computations of Wind Turbine Wakes

Stefan Ivanell, Jens N. Sørensen and Dan Henningson

48.1 Numerical Method

Numerical simulations using CFD are performed for wind turbine applications. The aim of the project is to get a better understanding of wake behaviour that cannot be obtained by standard industrial design codes for wind power applications. Such codes are based on the Blade Element Momentum (BEM) technique, extended with a number of empirical corrections that are not entirely based on physical flow features. The importance of accurate design models also increases as the turbines become larger. Therefore, the research is today undergoing a shift toward more fundamental approaches, aiming at understanding basic aerodynamic mechanisms.

Theoretically, the bound circulation on the blades is equal to the circulation behind the rotor, i.e. in the wake. For inviscid flows, the sum of the circulation of the tip and the root vortex should ideally be zero. However, this is not entirely correct for viscous flow. The tip and root vortex do, however, both for inviscid and viscous flows, have different sense of rotation, i.e. different signs of the circulation. A steep decline of circulation toward the tip will lead to a rapid concentration of the vortex at the tip (occurring a few chord lengths behind the tip). The sign of the circulation gradient along the blade will also determine the sense of rotation of the vortex behind the blade.

The simulations are performed using the CFD program "EllipSys3D" developed at DTU (The Technical University of Denmark) and Risø. The so-called Actuator Line Method is used, where the blade is represented by a line instead of a large number of panels. The forces on that line are introduced by using tabulated aerodynamic coefficients. In this way, computer resources are used more efficiently since the number of node points locally around the blade is decreased, and they are instead concentrated in the wake behind the blades. The actuator line method was introduced by Sørensen and Shen [1] and later implemented into the EllipSys3D code by Mikkelsen [2]. EllipSys3D is a general purpose 3D solver developed by Sørensen [3] and Michelsen [4,5].

260   S. Ivanell et al.

The presence of the rotor is modelled through body forces, determined from local flow and airfoil data. The Navies–Stokes equations are formulated as

$$\frac{\partial u_i}{\partial t} + u_j \frac{\partial u_i}{\partial x_j} = -\frac{1}{\rho}\frac{\partial p}{\partial x_i} + f_{\text{body},i} + \nu \frac{\partial u_j^2}{\partial x_j^2} \qquad (48.1)$$

where f_{body} represents the force extraction from the blades.

To avoid numerical instabilities, the forces are distributed among neighbouring node points using a Gaussian distribution.

48.2 Simulation

The result from the CFD simulation is evaluated and special attention is paid to the circulation and the position of vortices. From the evaluation, it will hopefully be possible to improve the engineering methods and base them, to a greater extent, on physical features instead of empirical corrections. Since the actuator line method uses tabulated airfoil data from measurements, data with good quality must be used. Data from the turbine Tjaereborg has been used for all simulations in this project [6]. The performed simulations are carried out in cylindrical coordinates and the mesh is reduced to a 120° slice with periodic boundary conditions to account for the rest of the required cylinder volume (see Fig. 48.1). The simulations are performed for steady conditions only.

The mesh was created as a 5-block mesh to be able to capture large gradients in the wake. To evaluate the sensitivity of the grid four different meshes with different resolutions were constructed; 48, 64, 80, and 96 nodes on each

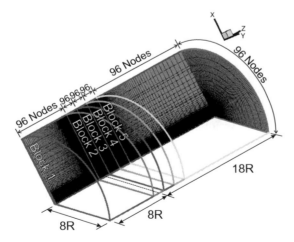

Fig. 48.1. The 5-block mesh and the distribution of the nodes. There are 96 nodes on each block side. In total this mesh contains $96^3 \times 5 \approx 4.4 \times 10^6$ nodes

48 Numerical Computations of Wind Turbine Wakes 261

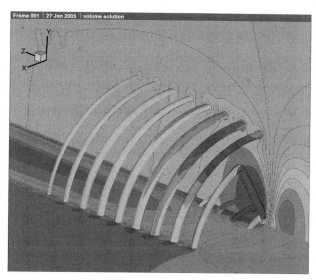

Fig. 48.2. $x = 0$-plane, pressure distribution; $y = 0$-plane, streamwise velocity; iso surface, constant vorticity with a surface of a contour pressure distribution

block side. The total number of node points are then: $5.5 \times 10^5, 1.3 \times 10^6, 2.6 \times 10^6$, and 4.4×10^6.

The sensitivity of the Gaussian smearing and the Reynolds number has also been studied.

An evaluation method to extract values of the circulation from the wake flow field is developed. When the circulation is evaluated in the wake, an integration is performed around a loop, enclosing the vortex. Each vortex is evaluated in terms of its circulation in a plane perpendicular to the turbine disc. The vortex created at the tip, or at least close to the tip, is evaluated every 30° behind the blade. The root vortex is evaluated in the same manner but since the vortices tend to be smeared out further downstream, because of diffusion, it is more difficult to evaluate the circulation further downstream in this case. The circulation is integrated at a specific value of the vorticity.

The sensitivity of the circulation with respect to the grid resolution has been evaluated. The evaluation was performed with three different grid sizes, 64, 80, and 96 node points at each side of the blocks. The solution converges with greater grid size.

Some wiggles appear at some distance downstream and can probably be explained by a too few number of grid points when the circulation is integrated and because of increasing smearing of the cores further downstream. When using a finer grid, the vortices become more concentrated and the integration can be performed further downstream without large fluctuations in the result.

262   S. Ivanell et al.

Assuming that all wake vorticity is included in the tip and the root vortex, the circulation at tip and root should approximately correspond to the maximum bound circulation of the blade. This is of course an idealization, since the roll-up of the vortex sheets is expected to form a concentrated vorticity distribution in the far wake [7]. The computed circulation at the blades is given in Fig. 48.3.

Figure 48.3 shows some wiggles that start at about 500° behind the blades for the root vortex and about 800° behind the blade for the tip vortex. These are most likely due to integration errors when integrating the circulation. At these positions the vortex core starts to be smeared out and it therefore becomes difficult to identify a good integration path.

The results, however, correspond fairly well to classical theories from Helmholtz. The tip vortex leaves the blade with a circulation value close to the maximum bound circulation at the blades. The root vortex leaves the blade with a much lower value of the circulation, but it increases rapidly and reaches values at the same order as the bound circulation. The reason why the root vortex is smeared out earlier than the tip vortex is partly because the vortex cores are closer at the root and partly because of the radial gradient of the circulation. The result is in good agreement with classical theorems from Helmholtz, from which it follows that the wake tip vortex has the same circulation as the maximum value of the bound circulation on the blade. A likely explanation for this good agreement is that the symmetry of the flow domain prohibits the roll-up of the vortex sheet from the blades.

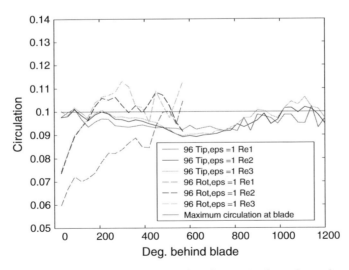

Fig. 48.3. The figure shows the circulation distribution in the wake at three different Reynolds number compared with the maximum circulation at the blade [8]

References

1. Sørensen J N, Shen W Z, Numerical Modeling of Wind Turbine Wakes, Journal of Fluid Engineering, vol. 124, June 2002
2. R. Mikkelsen (2003) Actuator Disc Methods Applied to Wind Turbines. PhD Thesis, DTU, Lyngby, MEK-FM-PHD 2003-02
3. Sørensen N N, General perpose flow solver applied to flow over hills, PhD Thesis, Risø National Laboratory, Roskilde, Denamark, 1995
4. Michelsen J A, Basis3D – a platform for development of multiblock PDE solvers, Report AFM 92-06, Dept. of Fluid Mechanics, DTU, 1992
5. Michelsen J A, Block structured multigrid solution of 2D and 3D elliptic PDE's, Report AFM 94-06, Dept. of Fluid Mechanics, DTU, 1994
6. Tjaereborg Data, http://www.afm.dtu.dk/wind/tjar.html
7. Sørensen J N, Okulev V L, Modelling of Far Wake Behind Wind Turbine, Proceeding of the Euromech Colloquium 464b – Wind Energy, Springer, 2006
8. S. Ivanell (2005) Numerical Computations of Wind Turbine Wakes, Lic. Thesis, Royal Institute of Technology, Stockholm, ISRN/KTH/MEK/TR/–05/10–SE

49

Modelling Wind Turbine Wakes with a Porosity Concept

Sandrine Aubrun

49.1 Introduction

The wind quality of a site is neither controllable nor improvable by wind energy specialists. The wind turbines arrangement in a wind farm is however designed by them before the implantation of a wind farm, in order to minimize turbine interactions. It is therefore of great interest to correctly assess these interactions. Since the field measurements are rare and difficult to interpret, wind turbine wakes and their interactions are usually treated with numerical models. An alternative to this is the physical modelling in wind tunnels. Indeed, its degree of modelling is lower than for numerical approaches.

Wind turbines are not plain obstacles and can be considered as porous elements, extracting kinetic energy from the flow, distorting streamlines and generating turbulence. The use of a porosity-drag approach, inspired from concepts developed to model the urban of forest canopy flows (numerical [1] and physical [2,3]) may be well adapted to simulate the wind turbine wake.

The present project is to model in a wind tunnel at a geometric scale of 1:400 wind turbines with metallic mesh discs to replicate the actuator discs. A parametric study of the flow field downstream of the wind turbine model has been performed in a homogeneous and turbulent approaching flow upon the disc size, the porosity level, the mesh size. Then the interaction between several porous discs located in a modelled atmospheric boundary layer was studied in order to model an offshore located wind farm.

49.2 Experimental Set-up

The wind turbines were modelled according to the actuator disc concept. The actuator disc extracts kinetic energy, generating a spreading of the stream tube, a velocity deficit downstream of the disc and an appearance of shear-generated turbulence. A porous disc produces exactly the same features [2]. Discs of 100 mm, 200 mm and 300 mm of diameter (D), made of metallic mesh,

266   S. Aubrun

Table 49.1. Metallic meshes properties

Mesh number	Mesh size (mm)	Wire diameter (mm)	Porosity level (%)
1	5	1.3	60
2	3.75	0.9	65
3	5.6	1.4	65
4	5.6	1.12	70
5	7.1	1.8	65
6	5	1.8	55
7	7.1 + 3.75	1.8 + 0.9	≤ 55

were used. Seven different meshes were tested (see Table 49.1). The discs were fixed to the floor with a vertical cylinder of diameter 0.05D, which is in agreement with standard wind tubine mast diameters.

Measurements were performed in the "Malavard" wind tunnel of the Laboratoire de Mécanique et d'Energétique. It is a close-circuit type with a test section of 2 m wide, 2 m high and 5 m long. A turbulence grid made of square tubes is located at the entrance of the test section. Its grid cell size is 100 mm and the tube cross-section is 25 mm. It generates a homogeneous and isotropic turbulence with a turbulence intensity of 3.5–4.5% in the used part of the test section. The first set of measurements is focused on homogeneous freestream conditions. In that case the porous disc is located in the middle of the test section. For the second set of experiments, the freestream conditions were representative of the offshore atmospheric boundary layer. The boundary layer which was developing on the wind tunnel floor (made of fairly rough wooden plates) in presence of the turbulence grid had the characteristics of the neutral offshore atmospheric boundary layer at a geometric scale of 1:400. The roughness length is $z_0 = 3$ mm in full scale, the power law exponent is $\alpha = 0.11$. The streamwise turbulence intensity at $z = 20$ m in full scale is $I_u = 7.3\%$. In that case, only the 100 mm-diameter discs made of mesh 6, combined with a 1D high mast, were used. At the chosen geometric scale, these discs replicate 40 m-diameter wind turbines.

The three components of velocity were measured with a Dantec triple hot-fiber probe connected to the Dantec StreamLine system. The probe was fixed on a 3D automated traverse.

49.3 Results for Homogeneous Freestream Conditions

A parametrical study was performed in a homogeneous freestream flow in order to define the influence of the disc size, the porosity level, the mesh cell size and the wire size on the flow. Mean velocity and turbulence intensity were measured for each case along the horizontal cross-section of the disc.

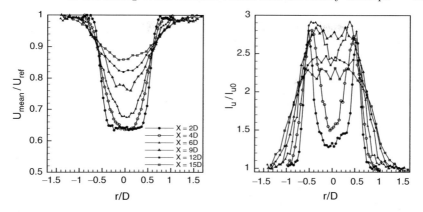

Fig. 49.1. Streamwise velocity and turbulence intensity downstream of a porous disc

Figure 49.1 presents results for a 100 mm-diameter disc made of mesh 6 at different downstream locations. The near-wake area, characterized by an annular shear layer and an annular turbulence intensity distribution, is well defined and ends up at $x/D = 4$, where the two shear layers start collapsing. The deduced axial flow induction factor is $a = 0.18$. The far wake starts at $x/D = 9$, from where the mean velocity and turbulence intensity profiles are self-similar. At $x/D = 9$, the turbulence intensity is locally three times higher than the freestream turbulence intensity I_{u0}.

Tests on the influence of the porosity level on the velocity and turbulence distributions show that one can totally control the velocity deficit, and so the shear-generated turbulence, using the appropriate mesh porosity level. Tests on the mesh cell size for a fixed porosity level (65%) illustrate that when the ratio between the mesh cell size and the disc diameter is smaller than 0.05, the velocity distribution is independent of the mesh. If this ratio is respected, it is also independent of the disc size.

49.4 Results for Shear Freestream Conditions

A wind farm was built with 9 square-arranged porous discs made of mesh 6 mounted on masts (Fig. 49.2). The masts spacing is $\Delta x = \Delta y = 3D$. Figure 49.3 shows the mean velocity distributions 3D downstream of the first, the second and the third row of porous discs. The black circle shows the disc circumference. Classical features of the wind turbine wakes are found [4]. As expected, the velocity deficit is located lower than the disc centre, due to the ground effect. The velocity deficit increases between the first and the second row, but stays constant between the second and the third one, although the wake extend is still growing. Figure 49.4 shows the turbulence intensity distributions at the same locations. It is worth to notice that, after three rows, the turbulence intensity can locally reach 26%, due to the combination of the

268   S. Aubrun

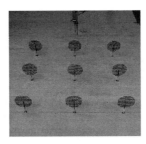

Fig. 49.2. The modelled wind farm

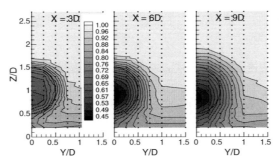

Fig. 49.3. dimensionless streamwise velocity, 3D downstream of the first, the second and the third row of discs

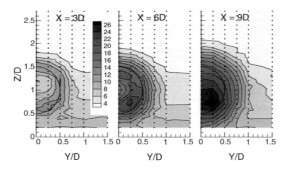

Fig. 49.4. Streamwise turbulence intensity, 3D downstream of the first, the second and the third row of porous discs

velocity fluctuations increase and the velocity deficit. The annular shape of the turbulence intensity distribution is still visible after the second row. Farther downstream, this shape is not predominant anymore. It is in agreement with the fact that no additional velocity deficit is generated at the third row.

49.5 Conclusion

The feasibility of this concept is proved and the parametrical study about the porous disc shows how the different disc features control the velocity deficit. Playing with the porosity level, any velocity deficit could be reproduced. Playing with the porous material homogeneity, one could reproduce even a non-uniform velocity deficit, which is more realistic. These experiments show also that the physical modelling is a good alternative and/or complement to the numerical modelling. For fundamental studies, it can help to better understand the interactions between wind turbine wakes and wind farms, to quantify the impact of wind farms on the atmospheric boundary layer, to validate numerical models and to develop the numerical porosity-drag concept. For applied studies, real situations can be reproduced in wind tunnel. Even complex terrain can be studied. The wind resource assessment for actual wind farms, as well as the wind farm impact on the site, can be performed.

References

1. Dupont S., Otte T.L., Ching J.K.S. (2004) Simulation of Meteorological Fields within and above Urban and Rural Canopies with a Mesoscale Model (MM5). Bound-Layer Metereol. 113/1, 111–158
2. Vermeulen P.E.J., Builtjes P.J.H. (1982) Turbulence measurements in simulated wind turbine clusters. Report 82-03003, TNO Division of Technology for Society
3. Aubrun S., Koppmann R., Leitl B., Mllmann-Coers M., Schaub A. (2005) Physical Modelling of a complex forest area in a wind tunnel – Comparison with field data. Agricul. Forest Metereol. 129, 121–135
4. Vermeer L.J., Sorensen J.N., Crespo A. (2003) Wind turbine wake aerodynamics. Progress in Aerospace Sciences 39: 467–510

50

Prediction of Wind Turbine Noise Generation and Propagation based on an Acoustic Analogy

Dragoş Moroianu and Laszlo Fuchs

The acoustical field behind a complete three dimensional wind turbine is considered numerically. Noise generated by the spatial velocity variation, force exerted by the blade on the fluid, and blade acceleration is taken into account. The sources are extracted from a detailed, instantaneous flow field which is computed using Large Eddy Simulations (LES). The propagation of the sound is calculated involving an acoustic analogy developed by Ffowcs Williams and Hawkings. It is found that the near field is dominated by the blade passage frequency. The results, also include ground effects (sound wave reflections) and the acoustical influence of a neighboring turbine.

50.1 Introduction

A very popular approach for aeroacoustic computations nowadays is a hybrid method comprising two steps: a flow solution is providing the noise sources, and then the acoustic waves generated by the noise sources are transported through the whole computational domain [1]. In this paper, an LES solution of an incompressible flow is used to extract the noise sources, and the Ffowcs-Williams and Hawkings acoustic analogy is used to compute the noise propagation.

50.2 Problem Definition

A three blade complete wind turbine, rotating with the angular speed of $60[rpm]$ in a wind of $10[m/s]$ is considered. The geometry is described in Fig. 50.1.

The boundary conditions used for this geometry are the following: inflow boundary condition on face $abcd$ - normal velocity with a magnitude of $10\,[m/s]$ and a turbulence intensity of $I = 4\%$ is imposed, outflow boundary

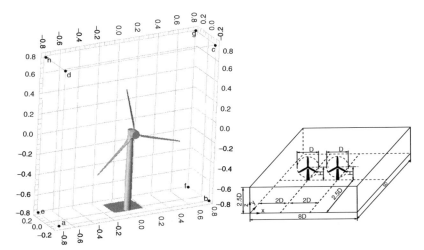

Fig. 50.1. Flow (left) and acoustical (right) computational domains

condition on face $efgh$ - constant pressure and a turbulence intensity of $I = 4\%$ is imposed, free stream boundary conditions on faces $bcgf$, $dcgh$ and $adhe$, the rest of the surfaces ($abfe$ and the surface defining the wind turbine) are wall boundary conditions.

The flow computational domain is included in a box with the following dimensions: $L_1 = L_2 = 1.54 \cdot D$ and $L_3 = 0.38 \cdot D$; where $D = 26[m]$ is the rotor diameter. The position of the rotor above the ground is $\frac{H}{D} = 0.8$. The cross section profile is from a NACA 4415 airfoil. The blade is twisted to keep a constant angle of attack of 7°. The spatial discretization of the domain was based on an estimation of the Taylor length scale outside the rotor and near the blade. The computational domain of the acoustical problem is much larger than the computational problem of the flow field, and the far-field of the acoustical field is depicted schematically in Fig. 50.1 (right). The boundary conditions that are set for the acoustical problem are as follows: on all boundaries, non-reflecting conditions are set, whereas on the ground, fully reflective conditions are imposed.

50.3 Results

50.3.1 Flow Computations

The flow contributes to the noise sources mainly at regions of large gradients in the flow as at the tip of the blade which is an important generator of shear. This vortices are formed due to the pressure difference between pressure and suction side of the blade and they are extended backward transported by the flow in a spiral mode, as seen in Fig. 50.2. They are interacting with the mast and with the vortices that are developing behind it.

50 Wind Turbine Noise 273

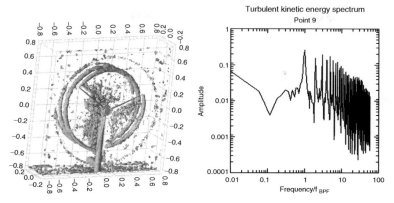

Fig. 50.2. Tip and hub vortices (left); Turbulent kinetic energy spectrum (right)

In Fig. 50.2 (left) can be noticed three hub vortices, twisted one over each other and spinning in opposite direction to the blades as it can be seen also experimentally [2].

Several vortical structures are developing behind the turbine starting from the mast. Point 9 is positioned in the tip vortex ($x/D = 0$; $y/D = -0.14$; $z/D = 0.49$) and it shows a spectrum (Fig. 50.2 (right)) which is dominated by the blade passage frequency.

50.3.2 Acoustic Computations

Several monitoring points are placed in the domain. $Y1(x/D = 2$; $y/D = 2.5$; $z/D = 1.3)$ is placed at the tip of the rotor and $Y6(x/D = 2$; $y/D = 3.75$; $z/D = 1.3)$ behind the turbine. $Z1(x/D = 2$; $y/D = 2.5$; $z/D = 0.1)$ and $Z6(x/D = 6.5$; $y/D = 2.5$; $z/D = 0.1)$ are placed at the foot of the mast and lateral to the turbine respectively. The acoustic density fluctuation ρ' decreases in amplitude with the distance from the rotor. The spectra of acoustic density fluctuation in Fig. 50.3 show that in the near field, the frequency range is broader and shrinks towards smaller values with the distance from the rotor. Except the blade passage frequency $f = 1$ and its harmonics, the spectra have a peak in the high frequency part (Fig. 50.3). This peak is shifted towards smaller values when the distance from the wind turbine is increased. This is an expected behavior and is caused by the production of the turbulence originating in the shear near the rotor blades which gets dissipated with the distance from its source.

Although the instantaneous acoustic density fluctuation field has a similar distribution from blade to blade, some differences can be noticed. This differences (Fig. 50.3 (below left)) are caused mainly by the influence of the mast over the flow. As seen in the same figure, the ground reflected waves interact with the propagating waves from the wind turbine, also the waves radiated from each neighboring turbine are interacting with each other.

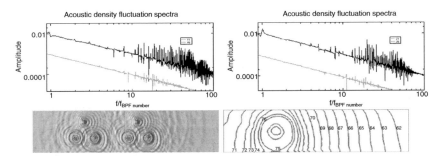

Fig. 50.3. Acoustic density fluctuation spectra (above left and right); Instantaneous acoustic density fluctuation (below left); Sound pressure level (below right)

The corresponding sound pressure level (SPL) field is depicted in Fig. 50.3 (below right). As expected, in the far field the SPL intensity decreases as if the wind turbine is a point source. The isocontours become almost spherical at a distance of about $3D$. The interferences from the ground in the far field are also weak and cannot be observed in the sound pressure levels.

50.3.3 Conclusions

The focus of the present work has been on the computation of the flow around a wind turbine as well as on the spreading of the sound that it generates. The vortex dynamics have been found to be in agreement with experiments. In order to compute the noise generated by the turbine, an acoustic analogy has been used. The acoustical field close to the wind turbine is dominated by the rotation frequency of the blades (BPF). In the far field the spectrum is influenced by the ground so different modes are stronger. There is a decay in the amplitude of the sound with the distance and far from the wind turbine, the acoustical waves are spread spherically as if they are coming from a point source. The influence of the mast in the process of generation and propagation of sound is not so strong. To asses the influence of terrain irregularities or air dissipation, further work is required.

References

1. Brentner, K. S., Farassat, F., "Modeling Aerodynamically Generated Sound of Helicopter Rotors", Progress in Aerospace Sciences 39, pp. 83–120, 2003
2. C.G. Helmis, K.H. Papadopoulos, D.N. Asimakopoulos, P.G. Papageorgas, "An experimental study of the near-wake structure of a wind turbine operating over complex terrain", Solar Energy, vol. 54, no. 6, pp. 413–428, 1995; 0038-092X(95)00009-7

51

Comparing WAsP and Fluent for Highly Complex Terrain Wind Prediction

D. Cabezón, A. Iniesta, E. Ferrer and I. Martí

51.1 Introduction

Assessing wind conditions on complex terrain has become a hard task as terrain complexity increases. That is why there is a need to extrapolate in a reliable manner some wind parameters that determine wind farms viability such as annual average wind speed at all hub heights as well as turbulence intensities.

The development of these tasks began in the early 1990s with the widely used linear model WAsP and WAsP Engineering especially designed for simple terrain with remarkable results on them but not so good on complex orographies. Simultaneously non-linearized Navier–Stokes solvers have been rapidly developed in the last decade through Computational Fluid Dynamics (CFD) codes allowing simulating atmospheric boundary layer flows over steep complex terrain more accurately reducing uncertainties.

This paper describes the features of these models by validating them through meteorological masts installed in a highly complex terrain. The study compares the results of the mentioned models in terms of wind speed and turbulence intensity.

51.2 Alaiz Test Site

Alaiz hill is 4 km long, 1,050 m high a.s.l. and is surrounded by a complex terrain associated to a ruggedness index (RIX) of 16%. The roughness level is high since the hill is covered by dense forests whereas the area upwind is clear without remarkable roughness elements.

Three meteorological masts located on the hill were employed in the study (Alaiz2, Alaiz3, and Alaiz6) forming a one year data base composed by hourly wind speed and wind direction values. Direction analysis afforded two main prevailing sectors at north and south.

276   D. Cabezón et al.

51.3 Description of the Models

51.3.1 Linear Models. WAsP 8.1 (Wind Atlas Analysis and Application Program) and WAsP Engineering 2.0

The linear model WAsP developed by Risoe allows simulating wind behaviour by obtaining the so-called geostrophic wind atlas regime taking into account the effects of terrain variation, surface roughness and nearby obstacles at a local mast. The model, as it occurs with other linear models, is limited to neutrally-stable wind flows over low, smooth hills with attached flows [1, 2].

The wind atlas offers the possibility to spatially extrapolate the wind statistics obtained at a certain meteorological mast to different hub heights at other locations. The program has been validated at different sites and widely used for assessing wind.

On the other hand, WAsP Engineering developed also by Risoe and introduced in 2001, simulates extreme wind, shear, flow angles, wind profiles and turbulence, being made as a complement of WAsP [3].

The purpose of WAsP Engineering is supporting the estimation of loads on wind turbines and other civil engineering structures in complex terrain.

51.3.2 Non Linear Models. Fluent 6.2

The Navier–Stokes solver Fluent 6.2 is one of the world's leading CFD commercial packages widely validated for a huge variety of flows supporting different mesh types. This non-linearized solver permits to recognise detached flows and to obtain iteratively the velocity magnitude and its components, the static pressure and the fields of turbulent kinetic energy (TKE) and turbulence dissipation rate through the K–ε model. In this study, wind is considered a 3D incompressible steady flow in which Coriolis force and heat effects have been omitted so neutral state of the atmosphere is considered [4].

51.4 Results

Validation was made by comparing the measuring campaign and the simulated wind speed and turbulence intensity from a horizontal and vertical extrapolation (for those values contained in the interval 350°–10°) between the lower level at one mast and the higher levels at the rest.

51.4.1 Wind Speed

The comparison shown in Table 51.1 indicates that CFD extrapolates wind speed between masts more accurately in almost all cases giving an average absolute error of 1.75% significantly less than the others: 5.67% for WAsP Eng and 5.41% for WAsP.

Table 51.1. Measurements and simulation results for wind speed and turbulence intensity

Input mast		Wind Speed			Turbulence Intensity		Output mast
		WAsP	WAsP Eng	Fluent	WAsP Eng	Fluent	
al2_20m	$\Delta V/\Delta TI$	1.10	1.22	1.09	0.77	1.01	WS $= 8.96\,\mathrm{ms}^{-1}$
WS $= 8.15\,\mathrm{ms}^{-1}$	al3_30m	8.96	9.98	8.92	8.40	11.01	TI $= 12.62\%$
TI $= 0.93\%$	Error (%)	0.02	11.41	-0.47	-33.43	-12.46	
	$\Delta V/\Delta TI$	1.26	1.23	1.10	0.78	0.81	WS $= 9.08\,\mathrm{ms}^{-1}$
	al3_40m	10.31	10.03	8.96	8.57	8.86	TI $= 11.68\%$
	Error (%)	13.55	10.46	-1.36	-26.68	-24.21	
	$\Delta V/\Delta TI$	1.27	1.23	1.11	0.80	0.34	WS $= 9.35\,\mathrm{ms}^{-1}$
	al3_55m	10.37	10.05	9.04	8.72	3.71	TI $= 10.67\%$
	Error (%)	10.93	7.51	-3.28	-18.28	-65.21	
	$\Delta V/\Delta TI$	1.15	1.13	1.04	0.90	0.30	WS $= 8.93\,\mathrm{ms}^{-1}$
	al6_40m	9.39	9.18	8.46	9.82	3.30	TI $= 8.22\%$
	Error (%)	5.17	2.82	-5.19	19.38	-59.82	
al3_30m	$\Delta V/\Delta TI$	0.84	0.86	0.93	1.28	0.52	WS $= 9.27\,\mathrm{ms}^{-1}$
WS $= 9.38\,\mathrm{ms}^{-1}$	Al2_40m	8.30	8.50	9.15	14.61	5.92	TI $= 7.55\%$
TI $= 11.38\%$	Error (%)	-10.51	-8.35	-1.37	93.50	-21.64	
	$\Delta V/\Delta TI$	0.92	0.92	0.95	1.17	0.30	WS $= 9.49\,\mathrm{ms}^{-1}$
	Al6_40m	9.02	9.06	9.36	13.35	3.41	TI $= 7.13\,\mathrm{ms}^{-1}$
	Error (%)	-4.91	-4.49	-1.37	87.32	-52.07	

Continued

Table 51.1. (*Continued*)

Input mast		Wind Speed			Turbulence Intensity		Output mast
		WAsP	**WAsP Eng**	**Fluent**	**WAsP Eng**	**Fluent**	
al6_20m	$\Delta V/\Delta TI$	0.95	0.96	1.00	1.12	0.67	WS = 9.12 ms^{-1}
WS = 9.19 ms^{-1}	Al2_40m	8.70	8.84	9.17	7.71	4.58	TI = 9.99%
TI = 6.89%	Error (%)	−4.55	−3.02	0.57	9.01	−35.20	
	$\Delta V/\Delta TI$	1.07	1.12	1.07	0.87	1.28	TI = 9.85 ms^{-1}
	al3_30m	9.85	10.26	9.88	6.01	8.82	TI = 8.70%
	Error (%)	0.03	4.19	0.31	−39.83	−11.75	
	$\Delta V/\Delta TI$	1.13	1.13	1.08	0.89	1.03	WS = 9.99 ms^{-1}
	al3_40m	10.35	10.41	9.92	6.12	7.09	TI = 8.70%
	Error (%)	3.60	4.20	−0.69	−29.58	−18.45	
	$\Delta V/\Delta TI$	1.13	1.12	1.09	0.90	0.43	WS = 10.3 ms^{-1}
	al3_55m	10.40	10.34	10.02	6.22	2.97	TI = 6.98%
	Error (%)	0.81	0.23	−2.90	−10.85	−57.44	
	Average	5.41	5.67	1.75	36.79	35.86	

Input mast/output mast = wind speed and turbulence intensities measured at input and output masts.
$\Delta V/\Delta TI$ = Change in wind speed/turbulence intensity between masts.

51.4.2 Turbulence Intensity

Table 51.1 also offers the comparison for turbulence intensity (TI) at the test site, which was carried out for WAsP Engineering 2.0 and Fluent 6.2.

As it is seen, both models obtained similar results when trying to extrapolate TI between masts. This result is also observed through the average absolute error: 36.79% for WAsP Engineering and 35.86% for Fluent.

The level of TI is not well captured by any of the studied models, being for CFD significantly lower than the measured values although the tendency was better modelled.

51.5 Conclusions

The analysis indicates that the non-linear solver Fluent 6.2 can simulate more accurately wind speed field for complex terrain than other wind flow models. Nevertheless, no conclusion could be extracted so far about which model explains better turbulence intensity.

Future works using CFD Fluent 6.2 will focus on testing more sophisticated turbulence models as well as on improving the distribution of surface roughness. Turbulence validation will also be done by means of more advanced wind sensors such as Lidar. On the other hand, grid independence studies will be carried out to reduce computing time.

These tasks will allow decreasing uncertainties in the near future when assessing wind farms power production in complex terrain.

References

1. Bowen and Mortensen (1996). Exploring the limits of WAsP. The wind atlas analysis and application program. Proceedings of the 1996 EWEC (Sweden)
2. Troen and Lundtang (1989). European Wind Atlas Risø (Denmark)
3. Mann et al. (2002). WAsP engineering 2000. Risø-R-1356(EN) pp. 90
4. Villanueva et al. (2004). Wind Resource Assessment in complex terrain using a CFD model. Proceedings of the 2004 EWEA Special Topic Conference. Delft (The Netherlands)
5. Ansys ICEM CFD 5.2 User Guide (2004). Ansys Inc
6. Fluent 6.2 User's Guide (2005). Fluent Inc
7. Richards and Hoxey (1993). Appropiate boundary conditions for computational wind engineering models using the K-Eps turbulence model, Journal of Wind Engineering and Industrial Aerodynamics, vol 46&47, pp.145–153

This study has been carried out thanks to Alaiz wind farm data provided by Acciona Energía.

52

Fatigue Assessment of Truss Joints Based on Local Approaches

H. Th. Beier, J. Lange and M. Vormwald

52.1 Introduction

A methodology to obtain Wöhler-curves is outlined in the present paper by applying it to truss joints with pre-cut gusset plates (PCGP-joint), Fig. 52.1. The approach itself is supposed to be generally applicable to new constructional details, of which the fatigue performance is certainly not yet classified.

An easier way to calculate the fatigue resistance of an unclassified structure is the use of the local stress approach. The local stress approach is exemplarily applied to discover the fatigue resistance of a critical detail of the "GROße WIndkraft ANlage" (great wind energy converter) GROWIAN, Fig. 52.2.

52.2 Concepts

A way to reduce the number of tests for deriving a Wöhler-curve is the additional use of appropriate and well established calculational models. If the calculation matches the test results, a number of further tests as far as parameter studies are concerned can then be done virtually by computing the fatigue resistance.

The proposed calculational concept takes into account all the three stages of fatigue life, fatigue crack initiation, crack growth and fracture separately. The amount of input data is small and the data can be found in the literature. The concept is in principle not limited to a certain group of structural details. It is evaluated for metallic materials.

The calculation of the fatigue endurance limit for welded structures can even be done by using the local stress approach and subsequently adding the Wöhler-curve with a Wöhler-curve exponent for welded structures, $m \approx 3$. The method does not take into account the three mentioned stages separately.

Fig. 52.1. PCGP-joint

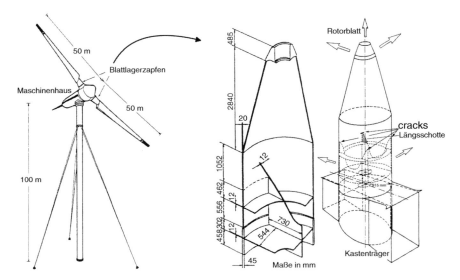

Fig. 52.2. Wind energy converter GROWIAN and plate-stiffener-connection

52.2.1 Fatigue Tests

The fatigue tests should give information about the location of the crack initiation and the corresponding number of cycles, the crack growth behaviour, the slope of the Wöhler-curve, the scatter of the test results and the FAT-class. The information is then used to verify the calculational fatigue resistance.

Information about the scatter can be found in the literature. Because the reported scatter takes a great number of different test situations into account, the standard deviation will be generally greater than by the own tests. In this case the literature values should be used for a statistical evaluation.

52.2.2 Crack Initiation with Local Strain Approach

The local strain approach relies on the assumption, that a certain cyclic stress–strain history at a local point in a structure will lead to the same damage as for

52 Fatigue Assessment of Truss Joints Based on Local Approaches 283

a standard unnotched specimen. Differences in surface, residual stresses and in size of structures and test specimens can be considered. Crack initiation is defined as a crack length of $c = 0.5\,\mathrm{mm}$, which is a crack depth $a \approx 0.25\,\mathrm{mm}$. As a result the number of cycles for crack initiation is obtained (Seeger [1]).

For the calculation of the local stress–strain history information must be given for:

– The cyclic stress–strain law, usually as Ramberg–Osgood law

$$\epsilon_a = \epsilon_{a,e} + \epsilon_{a,p} = \frac{\sigma_a}{E} + \left(\frac{\sigma_a}{K'}\right)^{1/n'}.$$ (52.1)

– The hysteretic behaviour, usually the Masing hypothesis, which means the doubling of the cyclic stress–strain law in stresses and strains.

The computation of the local stress–strain history can either be done by full elastic–plastic finite-element-calculations with true strains, or by linear-elastic calculations with subsequent approximation of the plastic behaviour, e.g. by Neuber's-rule [2]

$$\sigma\epsilon E = (K_t S)^2 = \sigma^2_{\mathrm{lin.elast.}}$$ (52.2)

The strain–life curve, usually tested with polished, unnotched specimens at a given stress ratio $R = -1$, can be described in the form of Manson, Coffin, Morrow [3–5]

$$\epsilon_a = \epsilon_{a,e} + \epsilon_{a,p} = \frac{\sigma'_f}{E}(2N)^b + \epsilon'_f(2N)^c.$$ (52.3)

The influence of mean stresses, i.e. the case $R \neq -1$, can be considered by damage parameters P, e.g. in the form of Smith et al. [6]

$$P_{\mathrm{SWT}} = \sqrt{(\sigma_a + \sigma_m)\epsilon_a E}.$$ (52.4)

The cyclic material characteristics can be taken from the uniform material law (UML) [7].

52.2.3 Crack Growth with Linear Elastic Fracture Mechanics

The calculation of crack propagation with linear elastic fracture mechanics starts at the crack depth $a = 0.25\,\mathrm{mm}$ from local strain approach. The crack depth dependent stress–intensity-factors $K_I(a)$ can be computed with FEM-software, e.g. FRANC2D/L [8]. A defined fracture criterion stops the calculation resulting in the number of cycles for crack growth.

Crack propagation can be calculated with the Paris-law [9]

$$\mathrm{d}a/\mathrm{d}N = C(\Delta K)^m.$$ (52.5)

Both material constants C and m can be found in literature.

284   H. Th. Beier et al.

52.2.4 Fracture Criterion

A fracture criterion (or limiting criterion) can be defined by fracture mechanics, for civil engineering structures, e.g. with DASt 009 [10] which is based on R6-method [11], by plastic collapse loads or deformation limits for ductile materials. Application of other criteria is possible.

For the PCGP-joint in [12] the demand of an overall elastic behaviour, similar to ASTM E647-99 [13], has been used as limiting criterion.

52.2.5 Endurance Limit with Local Stress Approach

An alternative approach which replaces Sects. 52.2.2–52.2.4 is the local stress approach. The approach in the form of Olivier [14] is based on one FAT class for all welds subjected to loads perpendicular to the weld direction. The method is established in IIW-recommendations [15]. The corresponding Wöhler-curve for an as-welded situation is given in [15] with FAT=225 and an exponent $m = 3$.

The approach is based on a calculation of effective notch stresses of welds by linear elastic calculation of stresses and a radius $r = 1\,\mathrm{mm}$ at the weld root and at the weld toe.

52.3 Examples

52.3.1 Truss-joint with Pre-cut Gusset Plates (PCGP-joint)

In order to obtain the fatigue resistance of the PCGP-joint the modern concept introduced in Sects. 52.2.1–52.2.4 has been used [12]: a minimised number of 15 tests have been performed on two types of normal scale specimens: (a) joints with sharp edges ($r = 2\,\mathrm{mm}$) and (b) with radii ($r = 10\,\mathrm{mm}$) at the free gap. The fatigue lives of these specimens have been calculated by the proposed concept. The Wöhler-curves of the tests and the calculations showed very good agreement, see Fig. 52.3 for the PCGP-joint-type with radii at the free gap (test series R2).

The verified calculation-model has then been used to perform an extensive parameter study to obtain the fatigue resistance of PCGP-joints typically used in structural steelwork. As a result the PCGP-joints with radii at the free gap can be classified in FAT 36 according to EC3 part 1 [16]. The joint version with the sharp edges can not be classified in a FAT-class and can therefore not be used in situations with cyclic loading.

52.3.2 Stiffener of the Great Wind Energy Converter GROWIAN

The great wind energy converter GROWIAN was build in 1983/1984 in northern Germany and worked for about 3 years in a "power supply system test mode". In 1987 operation was stopped due to severe cracks at the

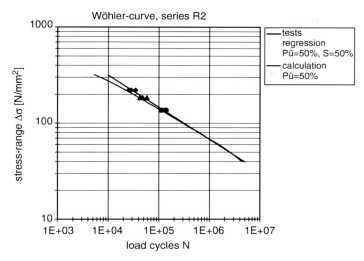

Fig. 52.3. Lives to fracture of PCGP-joint, series R2, tests and computed results

end of a welded plate-stiffener-connection after approximately 4×10^5 load cycles.

Köttgen et al. [17] answered the question about the fatigue resistance of the detail on the basis of nominal stresses, structural (geometric) stresses and effective notch stresses (local stress approach).

In fact, the GROWIAN-stiffener-detail has a special membrane and bending stress situation that can not be found in any nominal stresses based FAT catalogue. As a result, the "similar" situation of a long stiffener on a plate gave a sufficient fatigue resistance, which was obviously wrong.

Both the structural stress and the local stress approach showed an insufficient fatigue resistance for the detail. The structural stress approach predicted a failure at about 4×10^5 but the result depends on the way of extrapolating the stress field to get the structural stresses and on the selection of a suitable Wöhler-curve.

The local stress approach is independent of the overall geometry. The calculation predicted a failure after about 2×10^5 load cycles which was in good correspondence with the 4×10^5 cycles for failure.

52.4 Conclusion

Two methods to investigate the fatigue resistance of unclassified structural details have been introduced:

1. A modern method including a minimised number of fatigue tests together with the calculation of (a) crack initiation by the local strain approach

286   H. Th. Beier et al.

and (b) crack propagation with linear elastic fracture mechanics. The use of the method is shown by obtaining the Wöhler-curve of a truss joint with pre-cut gusset plates. The method can even be used to obtain the fatigue resistance of details under arbitrary load spectra.

2. The local stress approach to calculate the fatigue resistance of welded structures. The use of the method is shown by the evaluation of a stiffener-plate connection of the wind energy converter GROWIAN, which was damaged by cyclic loadings.

References

1. Seeger T (1996) Grundlagen für Betriebsfestigkeitsnachweise. In: Stahlbau Handbuch Teil 1 B. Stahlbau Verlagsgesellschaft, Köln
2. Neuber H (1968) Über die Berücksichtigung der Spannungskonzentration bei Festigkeitsberechnungen. Konstruktion 20:245–251
3. Manson S (1965) Fatigue: a complex subject – some simple approximations. Experimental Mechanics 5:193–226
4. Coffin LF (1954) A Study of the Effects of cyclic thermal stresses on a ductile metal. Transactions of ASME 76:931–950
5. Morrow JD (1965) Cyclic plastic strain energy and fatigue of metals. ASTM STP 378:45–87
6. Smith AN, Watson P, Topper H (1970) A stress function for the fatigue of materials. Internat. Journal of Materials 5:767–778
7. Bäumel A (1991) Experimentelle und numerische Untersuchung der Schwingfestigkeit randschichtverfestigter eigenspannungsbehafteter Bauteile. Dissertation, Technische Universität Darmstadt
8. James, MA (1998) A Plane Stress Finite Element Model for Elastic–Plastic Mode I/II Crack Growth. Dissertation, Kansas State University
9. Paris PC, Erdogan A (1963) Critical analysis of crack propagation law. Journal of Basic Engineering Transactions, ASME Series D 98:459
10. DASt Richtlinie 009 (2005) Stahlsortenauswahl für geschweite Stahlbauten. Stahlbau Verlags- und Service GmbH, Düsseldorf
11. R6 Revision 4 (2001) Assessment of the Integrity of Structures Containing Defects. British Energy Generation Ltd, Barnwood, Gloucester
12. Beier HTh (2003) Experimentelle und numerische Untersuchungen zum Schwingfestigkeitsverhalten von ausgeklinkten Knotenblechen in Fachwerkträgern. Dissertation, Technische Universität Darmstadt
13. ASTM-Standard E 647-99 (1999) Standard Test Method for Measurement of Fatigue Crack Grow Rates. American Society for Testing and Materials
14. Olivier R (2000) Experimentelle und numerische Untersuchungen zur Bemessung schwingbeanspruchter Schweißverbindungen auf der Grundlage örtlicher Beanspruchungen. Dissertation, Technische Universität Darmstadt
15. Hobbacher A (1996) Fatigue design of welded joints and components. Recommendations of IIW. Abington Publishing, Cambridge England
16. Eurocode 3 Part 1–9 (2000) Fatigue strength of steel structures. CEN, Brussel
17. Köttgen VB, Olivier R, Seeger T (1993) Der Schaden an der Großen Windkraftanlage GROWIAN. Konstruktion 45:1–9

53
Advances in Offshore Wind Technology

Marc Seidel and Jens Gößwein

53.1 Introduction

The new generation of large wind turbines, which is specifically designed for Offshore application, is now in the prototyping phase. The currently largest wind turbine REpower 5M is in operation since autumn 2004. First offshore prototypes of this type shall be installed in 2006. This paper focuses on the mechanical and structural issues related to large Offshore Wind turbines. Other challenging areas – like electrical engineering, financing or operation and maintenance – are not discussed.

53.2 Wind Turbine Technology

Due to the demanding offshore environment with a reduced accessibility, the main emphasis in the development is an economic design of the turbine with a high reliability and a high availability. The REpower 5M is based on proven technology and can be regarded as a logical evolutionary step compared to the smaller turbines. This is a significant difference compared to other large wind turbines which are based on new concepts but does not have a proven track record.

The REpower 5M is equipped with a three-bladed rotor with pitch control. Rated power is 5,000 kW (5 MW) with a rotor diameter of 126 m. The prototype has been installed on a steel tower with a hub height of 120 m. The turbine has been designed for strong wind sites with an annual mean wind speed of $10.5\,\mathrm{ms}^{-1}$ at hub height.

The main technical features of the REpower 5M can be summarized as follows (see also [1]):

- Modular drive train with flanges
- Double drive train bearings (CARB^{TM} and spherical roller bearing)
- On board crane for Operation and Maintenance (Capacity: 3.5 t)

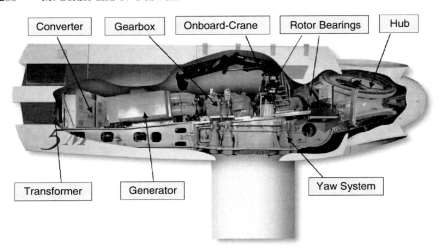

Fig. 53.1. Arrangement of main components

- Service friendly design
- Power electronics and transformer in the nacelle

The rotor blades have been developed in cooperation with LM Glasfiber with the largest rotor diameter ever designed. The blade structure of the LM 61.5 is optimized such that a low blade weight is achieved, in combination with high static permissible stress and high energy yield.

The main, or rotor shaft is supported by two bearings. The first, non locating bearing is a so-called CARB^{TM}-bearing, which can handle much higher dislocations compared to a conventional spherical roller bearing. The second, locating bearing is a spherical roller bearing, which protects the gearbox from external loads (other than the torque) to a large extent. Thus the double bearings will increase the safe life of the gearbox.

The gearbox is a planetary gear with one spur wheel. It is equipped with a refined system for oil filtration and cooling. The oil is filtrated down to 6 μm and the cooling system prevents high oil temperatures. Thanks to the double bearing of the main shaft and a robust rotor lock, the gearbox can be disassembled without taking the rotor down.

A double fed inducing generator is used, together with an IGBT inverter. Four water cooled inverters are used with a redundant combination of every pair of two modules. This means, that on failure of one module (or pair), the turbine can still operate at reduced power (up to 3 MVA).

The GFRP nacelle panel and spinner consist of several parts and are protected by a sophisticated lightning protection system.

The simulations of the 5M are extensively compared with measurements. Results so far show a very good agreement between the simulations with Flex 5 and the measurements (see Fig. 53.2).

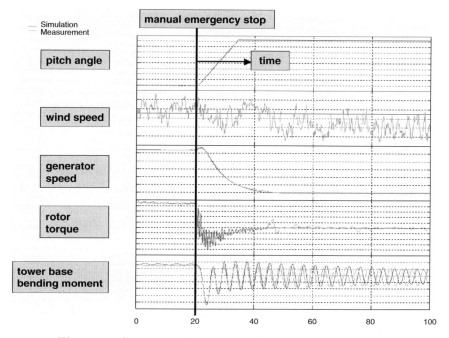

Fig. 53.2. Comparison between simulation and measurement

53.3 Substructure Technology

One of the main drivers for an economic installation is the substructure design. Many developments on substructure concepts are ongoing and the industry is clearly moving towards more competitive concepts. REpower has worked extensively on design methods for the combined wind and wave loading and several substructures have been evaluated from technical and economical points of view.

53.3.1 Design Methodologies

So far a substitute Monopile is used to represent complex substructures for simulation of the turbines [2]. Although this "semi-integrated" methodology is workable for many substructures, there are some drawbacks:

– Stiffness representation with a Monopile is not adequate for all kinds of substructures. This leads to significant differences in overall stiffness and thus eigenfrequencies, which are an important parameter in the transient calculations.
– Wave loading is calculated for a straight vertical member, so that the positive effect of distributed members in wave direction is not taken into

account. Furthermore, finding equivalent hydrodynamic coefficients to take account of many members proves to be difficult in many cases.
- The use of foundation models in two different programs, which need to be harmonized, e.g. in terms of wave loads, creates an additional interface with more possibilities for errors in the process.

The shortcomings of this approach lead to the development of a new solution with a higher degree of integration of the two programs used [3]. The characteristics of this integrated, sequential approach are:

- Two specialized programs – Flex 5 for the wind loads and ASAS(NL) for the wave loads – have been adapted to each other such that they can be used sequentially to obtain a solution for the integrated system.
- The general approach to achieve this has been based on the existing modelling in Flex 5 which uses generalized degrees of freedom to model the substructure. Six generalized degrees of freedom are used.
- Although the methodology is still an approximation, good results can be achieved for many structures. Good knowledge about the methodology and its limitations are however required for a safe and efficient design.

A flow chart of the methodology is shown in Fig. 53.3.

53.3.2 Substructure Concepts

So far, most of the existing Offshore turbines have been installed with Monopile or gravity base substructures. Due to the size of the 5M and the associated loading, Monopiles are getting quite large and thus other concepts may become more interesting, especially for water depths in excess of 10 m.

Extensive work has been done for the DOWNVInD project with two turbines for the demonstration phase which will be installed in 45 m water depth in 2006. Initially, Tripod foundations which had a weight of over 1,000 t were foreseen for this project. This selection has been revisited as REpower has worked closely with OWEC Tower – a Norwegian company with a background in offshore oil and gas which is now specializing in offshore wind turbines – and it was found that the OWEC Jacket Quattropod (OJQ) offers a much better chance for economic optimization due to lower weight of the substructure. The final design of this structure is currently ongoing and fabrication is expected to commence end of 2005.

Other concepts under evaluation include concrete substructures which are an interesting alternative due to the low material cost. Different concepts are offered on the market, primarily from large building contractors which aim at gaining market shares in this evolving market.

53.4 Installation Methods

Installation concepts are of course closely related to the structural concepts. Installations so far have been executed in a very similar way. Basically the

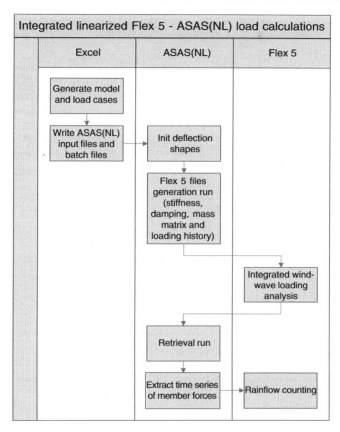

Fig. 53.3. Overview of calculation process

components substructure, tower, nacelle and rotor are installed sequentially, very much in the same way as onshore construction is performed.

The DOWNVInD project will feature significant innovative approaches. Due to the fact that conventional installation is not feasible as jack-ups for 45 m water depth are not available for economical conditions, an installation in two parts is foreseen. The OJQ will be installed and piled first and the full assembly of wind turbine, rotor and tubular tower will be installed as one complete unit subsequently. A detailed investigation on how and if this approach can be made to work is currently underway.

References

1. Schubert, M.; Gößwein, J.: Aufstellung und erste Betriebserfahrungen der weltweit größten Windenergieanlage REpower 5M. Offshore-Symposium GL-Windenergie. Hamburg 2005

2. Seidel, M.; v. Mutius, M.; Steudel, D.: Design and load calculations for Offshore foundations of a 5 MW turbine. Conference Proceedings DEWEK 2004. Wilhelmshaven: DEWI 2004
3. Seidel, M.; v. Mutius, M.; Rix, P.; Steudel, D.: Integrated analysis of wind and wave loading for complex support structures of Offshore Wind Turbines. Conference Proceedings Offshore Wind 2005: Copenhagen 2005

54

Benefits of Fatigue Assessment with Local Concepts

P. Schaumann and F. Wilke

54.1 Introduction

Support structures of offshore wind energy converters for medium to high water depths are supposedly carried out as braced or lattice constructions. Due to the combined dynamic loads from wind and waves the fatigue design is of special interest. The large variety in type and dimensions of tubular joints requires a fatigue design with local approaches.

54.2 Applied Local Concepts

The authors gave an overview of expected types of support structures in [1]. State-of-the-art in current offshore standards is the structural stress approach. More sophisticated concepts like the notch stress approach and notch strain approach [2], here referred to as "local approaches", require detailed FE modelling of the mesoscale weld shape.

The applied notch stress approach is based on the recommendations in [3], but with a modified slope of the S–N curve with $m = 5$ for $N_\mathrm{D} > 5 \times 10^6$ and without a cut-off limit. This modification fits well to the results of [4] in the high-cycle fatigue range. A fictitious weld toe fillet with a radius ρ_f=1 mm is applied to the FE model to account for microstructural support. The notch strain approach uses stabilized cyclic material data according to the uniform material law (UML). For the material in the heat-affected zone (HAZ) an increased hardness is considered. To include the mean stress effects, the damage parameter given by

$$P_{\mathrm{SWT}} = \sqrt{(\sigma_\mathrm{a} + \sigma_\mathrm{m})\epsilon_\mathrm{a} E} \qquad (54.1)$$

has been used. Together with the equations by Manson, Coffin and Morrow this leads to a relevant P–N curve. As the shear strains are low for the analysed joints and loads, only strains and stresses normal to the weld toe are taken into account (critical plane approach).

As a benefit of both local concepts fatigue design can be carried out uncoupled from the standard S–N curves for given structural details and the application to any local weld geometry is possible. At the same time the number of influencing parameters increases which is shown in the following chapter.

54.3 Comparison of Fatigue Design for a Tripod

The bottom part of the reference structure and the most important data are shown in Fig. 54.1. For the scatter diagram of the chosen Baltic Sea location see [1]. The relatively small turbine leads to dominating fatigue damage from wave loading.

Figure 54.2 shows the resulting member S–N curve for a fictitious pulsating nominal stress in one brace of the top tripod node. According to [5] three different engineering approaches for the notch strain concept have been applied (1. Heat-affected zone material without residual stresses and $R = -1$; 2. heat-affected zone material with $R = 0$; 3. base material (BM) with residual stresses in the order of the yield strength and $R = -1$). The notch stress curve has been shifted to a probability of failure $P_f = 50\%$ ($\gamma = 1.38$) to provide a basis for comparison with the results obtained through the use of the UML.

In high cycle regions beyond $N = 2 \times 10^6$, interesting for offshore structures, the compliance is acceptable. This leads to the conclusion that the notch stress approach seems to provide cycles up to crack initiation for high plate thicknesses and notch dominated problems. Figure 54.2 also shows an obvious handicap of the notch strain approach. Besides the usual scatter in the local weld geometry it is hard to determine the parameters which govern the design without experiments. Because of the higher robustness regarding the influencing parameters the notch stress approach (NSA) is used for further parametric studies.

Figure 54.3 (left) shows the comparison of cycles to failure for a sinusoidal pressure of $5\,\text{N}\,\text{mm}^{-2}$ on one brace of the top tripod node for a large variety

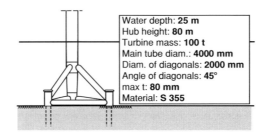

Fig. 54.1. Bottom part of the reference tripod

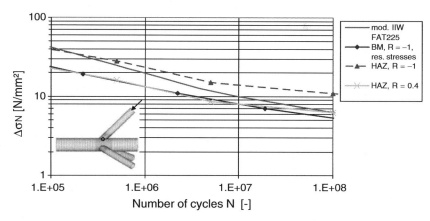

Fig. 54.2. Predicted crack initiation life for the top tripod tubular joint

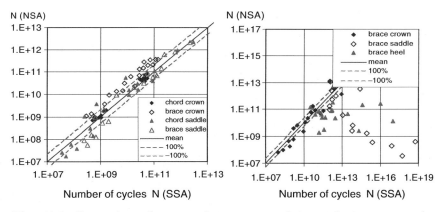

Fig. 54.3. Comparison of structural stress approach to notch stress approach

of dimensions and different crack locations. The compliance of the structural stress approach (SSA) with the NSA is excellent leading to a best-fit hot spot S–N curve FAT98 (compared to $100\,\mathrm{N\,mm^{-2}}$ as the reference value) and an average plate thickness exponent k of 0.245. The heel points are an exception because the underlying volume weld model with the gross section according to AWS results in too steep flank angles.

The benefits of the local approach are used for an application to the case of single-side welded joints for which insufficient test data is available. Here the comparison with a reduced hot-spot S–N curve as recommended by offshore standards shows an overestimation of lifetime by the SSA, which cannot be explained by the slight overestimation of strength reduction for short gaps mentioned in [1]. For the analysed tubular joints, local bending effects due to the weld shape stay uncovered by the hot-spot method, especially for the brace saddle locations.

54.4 Conclusion

Fatigue design is a multiparameter challenge. The number of parameters significantly increases the sensitivity of results when applying local concepts, in particular the notch strain approach. Therefore experimental verification of the theoretical analysis is indispensable but unlikely for the extreme large dimensions of the proposed structures. Nevertheless the two local concepts show their benefits if transfer to new structures or weld types is needed. For offshore structures with actions in the high-cycle region, the notch stress approach should be chosen. The application of the notch stress approach has shown a very good compliance with the structural stress approach for the weld toe locations, but great differences in fatigue strength for the roots. Here the design rules for single-side welded joints have to be examined in more detail with fracture mechanics.

References

1. Schaumann P, Wilke F (2005) Current developments of support structures for wind turbines in offshore environment. In: Shen et al. (eds.) Advances in Steel Structures, Elsevier, Amsterdam, pp. 1107–1114
2. Radaj D, Sonsino CM (1998) Fatigue Assessment of Welded Joints by Local Approaches. Abbington, Cambridge
3. Hobbacher A (1996) Fatigue design of welded joints and components. IIW Doc XIII-1539-96, Abbington, Cambridge
4. Olivier R, Köttgen VB, Seeger T(1994) Schweißverbindungen II–Untersuchungen zur Einbindung eines eines neuartigen Zeit- und Dauerfestigkeitsnachweises von Schweißverbindungen in Regelwerke. Forschungskuratorium Maschinenbau Heft 180, Frankfurt
5. Radaj D, Sonsino CM, Flade D (1998) Prediction of service fatigue strength of a welded tubular joint on the basis of the notch strain approach. Int. J Fatigue 20(6): 471–480

55

Extension of Life Time of Welded Fatigue Loaded Structures

Thomas Ummenhofer, Imke Weich and Thomas Nitschke-Pagel

Summary. At the University of Braunschweig the effects of ultrasonic peening (UP) and shot peening on the fatigue life and fatigue strength of new and existing wind energy plants have been investigated. The test results demonstrate that UP is an effective and practical means to extend their service life.

55.1 Introduction

The design of wind energy converters is significantly defined by fatigue. In the coming years a large number of converters will reach their designed service life of 20 years, so that methods to increase their fatigue life are required. At the University of Braunschweig research has been initiated on the application of ultrasonic peening and shot peening to increase the fatigue life of new and existing wind energy plants.

55.2 Background

55.2.1 Weld Improvement Methods

The effectiveness of weld improvement methods to increase the fatigue life has been investigated in several studies. Methods known are local or overall grinding, burr grinding, TIG-dressing, needle peening, hammer peening, shot peening, UP or ultrasonic impact treatment (UIT). Methods which combine geometric and mechanical effects show the best fatigue strength improvement. A large amount of data from literature show that S–N curves of improved specimens run flatter than the S–N curves of untreated specimens [1]. So far, only few empirical investigations on the effectiveness of UP on the increase of the preloaded specimens are available [2].

55.3 Experimental Studies

55.3.1 Testing Parameters

In this study MAG-welded flange segments (S355J2G3) directly attached to a flat web, by a double bevel seam have been investigated. The two half segments were connected by one bolt. Additionally a butt weld has been incorporated into the web. The specimens have been clean blasted (SA 2 1/2). For more details see [3, 4].

55.4 Results

55.4.1 Initial State of the Fatigue Test Samples

To characterize the surface changes caused by the different improvement methods micrographs and profiles measured by laser triangulation sensors of the weld toe of a butt welded joint in sand blasted, shot peened and ultrasonic peened condition have been compared. The graphs show that the peening treatment leads to a notch radius, directly connected with the radius of the tip of the used peening tool used. However, in the transition zone between the peened toe and the adjacent base material respectively the weld metal the material is raised causing small sharp geometrical notches (Fig 55.1) so that the weld toe profile is not always improved by the peening treatment. Transverse residual stress profiles and profiles of the integral width of the measured diffraction lines, both measured by X-ray diffraction, have been analysed and compared to an as-welded state observed after an electrochemical removal of a 0.3 mm thick surface layer (Fig 55.2). Whereas shot peening and clean blasting lead to an homogeneous cold hardening effect without significant peaks at the weld toe or in the weld seam the hardening effect of the UP is limited to the region of the treated weld toe. The residual stress distribution shows that only shot peening leads to an almost uniform state of

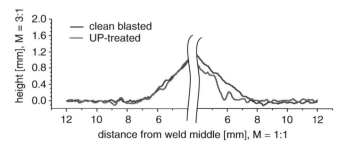

Fig. 55.1. Surface profiles in the weld toe region of a specimen before and after UP-treatment

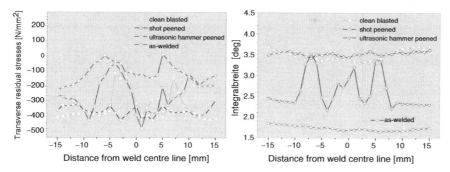

Fig. 55.2. Near surface hardness profiles after clean blasting (left side) and after UP-treatment (right side)

compressive residual stresses of 300–400 N mm^{-2}. The undefined clean blasting process also produces compressive residual stresses, however, with more scattering values. The UP also induces compressive residual stresses which are as high as for shot peening but limited to the treated zones at the weld toe. The magnitude of the compressive residual stresses in the treated zones are non-uniform and scatter, probably caused by the manual application of this treatment method. For more details see [3, 4].

55.4.2 Results of the Fatigue Tests

The results show that the fatigue life of the notch details could be highly increased using shot peening and UP. The fatigue cracks of the untreated specimens start basically in the double bevel seam. Regarding the calculated, nearly similar stresses in the butt welds, a similar fatigue life as for the double bevel seam can be predicted for those. The fatigue cracks of the UP-treated specimens occur in the middle of the treated zone or at the edges of the butt welds. These results correspond well to the measured sharp burrs at the edges of the dents of the impacts. The fatigue life could be increased by a factor around four compared to the untreated specimens. Due to bending stresses in the double bevel seam caused by the flange connection the influence of surface compressive stresses on the effective stresses are even higher so that the crack initiation shift from the double bevel seam to the butt weld. The increase of the fatigue life despite sharp notches confirms the beneficial influence of the surface hardening and the compressive residual stresses. Preloaded specimens which have been loaded up to the design life according EC3 and later UP-treated reach a mean fatigue life similar to immediately treated ones. The fatigue life of the shot peened specimens has also been highly increased, some specimens even fail in the base material. This confirms that the increase of the fatigue life is caused by the introduction of residual stresses. Figure 55.3 shows the S–N curves for the tested notch details, based on an assumed value of the slope m of 3 for the as-welded condition and 5.6 for the treated

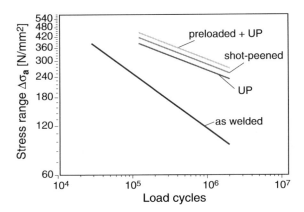

Fig. 55.3. S–N curves of the different treated notch details

specimens based on results of other studies [1]. The results of the preloaded and subsequently UP-treated specimens are even better than the results of the immediate treated ones but it has to be taken into consideration that the UP treatment has been slightly optimized. For more details see [3,4].

55.5 Conclusions

Test results show that the application of shot peening and UP on fatigue loaded butt welds and double bevel seams produce a high enhancement of the fatigue strength or life time. Ultrasonic methods proved to be simply manageable preventive or repair methods with similar beneficial effects as shot peening, also increasing the fatigue life of preloaded notch details. It appeared that the positive effects are based on strain hardening connected with positive residual compressive stresses.

References

1. Manteghi S., Maddox S. J. (2004) Methods for the Fatigue Life Improvement of Welded Joints in Medium and High Strength Steels. Proceedings of the International Institute of Welding, IIW Dokument XIII-2006-04
2. Garf E. F. et al. (2001) Assessment of Fatigue Life of Tubular Connections Subjected to Ultrasonic Peening Treatment. The Paton Welding Journal, Vol. 2: 12–15
3. Ummenhofer T., Weich I., Nitschke-Pagel T. (2005) Lebens- und Restlebensdauerverlängerung geschweißter Windenergieanlagentürme und anderer Stahlkonstruktionen durch Schweißnahtnachbehandlung. Stahlbau 74 H.8: 412–422
4. Ummenhofer T., Weich I., Nitschke-Pagel T. (2005) Extension of Life Time of Welded Dynamic loaded Structures. IIW Document XIII-2085-05

56

Damage Detection on Structures of Offshore Wind Turbines using Multiparameter Eigenvalues

Johannes Reetz

56.1 Introduction

An efficient and reliable use of offshore wind-energy requires a condition controlled maintenance. Thereby information about the state of damage and the residual load carrying capacity of the structure has to be provided by means of condition monitoring systems (CMS). Such a system is not yet available. In the following, a new method is demonstrated treating this problem. The method is based upon the fact that damages influence the dynamic behavior of elasto mechanical structures. Turek et al. [1] show that with the aid of a validated finite element model damage detection can be made by means of measured modal quantities. However, their method does not provide information about quantification and localisation of damage. By formulating the identification problem as a multiparameter eigenvalue problem using the finite element model with parametrised model matrices, damage can be quantified and localised [2].

56.2 The Multiparameter Eigenvalue Method

Starting from the known linear elastic and lightly damped supporting structure the first step is the approximation as a finite element model. Thereby the model has to be validated with the real structure. It is advantageous to apply pre-existing knowledge about structural area or parts, e.g. welded or bolted joints or the splash zone, which are prone to damage. For this purpose the portions of the model matrices – particularly of the stiffness matrix – associated with these details are parametrised. Herewith, the mathematical model characterising the dynamic behavior is given. On the other hand, the information about the current dynamic behavior of the potentially damaged structure is provided by the modal quantities. The modal quantities which are most exactly measurable are the eigenfrequencies. These can be measured by means of ambient excitation. The identification should provide the

302   J. Reetz

information about the current stiffness parameters by means of the given model and the measured eigenfrequencies. Due to the formulation as multi-parameter eigenvalue problem the ill-posed problem is transformed into common linear eigenvalue problems. The last step is the diagnostics. Thereby the interpretation of parameters provides detection, quantification and localisation of the damage. To simplify matters the equations hold for a number of two parameters without loss of generality. The problem of identification of the stiffness parameters by means of measured eigenfrequencies is

$$\begin{cases} \left(-\hat{\omega}_1 \mathbf{M} + a_1 \mathbf{K}_1 + a_2 \mathbf{K}_2\right) \mathbf{g}_1 = \mathbf{0}, \\ \left(-\hat{\omega}_2 \mathbf{M} + a_1 \mathbf{K}_1 + a_2 \mathbf{K}_2\right) \mathbf{g}_2 = \mathbf{0}. \end{cases} \tag{56.1}$$

These eigenvalue equations comprise the known model matrices $\mathbf{K}_i$ and $\mathbf{M}_i$, the known measured eigenfrequencies $\hat{\omega}_i$, the unknown eigenvectors $\mathbf{g}_i$ and the unknown stiffness parameters a_i, which have to be identified. The same elements in a generalised notation are given by

$$\begin{cases} \left(\lambda_0 \mathbf{A}_{10} + \lambda_1 \mathbf{A}_{11} + \lambda_2 \mathbf{A}_{12}\right) \mathbf{g}_1 = \mathbf{B}_1 \mathbf{g}_1 = \mathbf{0}, \\ \left(\lambda_0 \mathbf{A}_{20} + \lambda_1 \mathbf{A}_{21} + \lambda_2 \mathbf{A}_{22}\right) \mathbf{g}_2 = \mathbf{B}_2 \mathbf{g}_2 = \mathbf{0} \end{cases} \tag{56.2}$$

with the stiffness parameters $\lambda_i = \lambda_0 a_i$ and the model matrices $\mathbf{A}_{ik}$. Every row is posed in the space of physical coordinates. There is a dependence of the required information on the number of degrees of freedom (DOF) of the model and hence a lack of information. This problem is ill-posed. But with the linear maps $\mathbf{B}$ one can define linear induced maps $\mathbf{B}^\dagger$ as follows:

$$\mathbf{B}^\dagger : R^3 \otimes R^3 \to R^3 \otimes R^3, \tag{56.3}$$

$$\mathbf{B}_1^\dagger : (\mathbf{g}_1 \otimes \mathbf{g}_2) \mapsto \mathbf{B}_1 \mathbf{g}_1 \otimes \mathbf{g}_2,$$

$$\mathbf{B}_2^\dagger : (\mathbf{g}_1 \otimes \mathbf{g}_2) \mapsto \mathbf{g}_1 \otimes \mathbf{B}_2 \mathbf{g}_2.$$

The maps $\mathbf{B}_1$ and $\mathbf{B}_2$ act on the element $\mathbf{g}_1$ and $\mathbf{g}_2$, respectively and the identity acts on the other elements. The application of these induced linear maps on the problem shows in (56.4) that if and only if the right-hand side holds, the left-hand side also holds:

$$\begin{cases} \mathbf{B}_1 \mathbf{g}_1 = 0 \\ \mathbf{B}_2 \mathbf{g}_2 = 0 \end{cases} \Leftrightarrow \begin{cases} \mathbf{B}_1^\dagger \left(\mathbf{g}_1 \otimes \mathbf{g}_2\right) = \mathbf{B}_1^\dagger \mathbf{f} = 0, \\ \mathbf{B}_2^\dagger \left(\mathbf{g}_1 \otimes \mathbf{g}_2\right) = \mathbf{B}_2^\dagger \mathbf{f} = 0. \end{cases} \tag{56.4}$$

The condition written in model matrices is

$$\sum_{s=0}^{2} \mathbf{A}_{rs}^\dagger \lambda_s \mathbf{f} = \sum_{s=0}^{2} \mathbf{A}_{rs}^\dagger \mathbf{f}_s = 0. \tag{56.5}$$

This problem is posed in the space of tensor products. By extension of the use of determinants for systems of linear equations one can apply the rules of determinants on this linear induced operators. Then one obtains compositions of maps (see (56.6)) called determinantal maps, i.e.

56 Damage Detection on Structures of Offshore Wind Turbines   303

Table 56.1. Identified stiffness parameters by means of measured (simulated) eigenfrequencies with an accuracy of three correct decimal places

Measured eigenfrequencies (Hz)	Determined stiffness parameters (–)
0.407	1.002
2.782	0.897
7.618	1.003

$$\boldsymbol{\Delta}_0 = \det \begin{pmatrix} \mathbf{A}_{11}^\dagger & \mathbf{A}_{12}^\dagger \\ \mathbf{A}_{21}^\dagger & \mathbf{A}_{22}^\dagger \end{pmatrix} = \mathbf{A}_{11}^\dagger \mathbf{A}_{22}^\dagger - \mathbf{A}_{21}^\dagger \mathbf{A}_{12}^\dagger. \tag{56.6}$$

It is shown in [3] that the development of determinants using Cramers Rule and the application of the cofactor to (56.5) lead to a problem which is equivalent to (56.1)

$$\boldsymbol{\Delta}_p \mathbf{f}_q - \boldsymbol{\Delta}_q \mathbf{f}_p = 0. \tag{56.7}$$

With a non-singular linear combination of the determinantal maps one obtains linear eigenvalue problems for the model parameters

$$\mathbf{f}_s = \boldsymbol{\Delta}^{-1} \boldsymbol{\Delta}_s \mathbf{f} = \lambda_s \mathbf{f}. \tag{56.8}$$

These problems are well-posed. There is no incompleteness of information because the number of required eigenfrequencies depends only on the number of stiffness parameters.

Numerical Example: In the current state of method validation, measurement data are not yet available. Therefore, artificial measurement data are generated by applying scatter to the exact eigenfrequencies as resulting from the analysis. It is supposed that the middle area of a cantilever beam model with three elements (denoted by one to three beginning from the bottom) and nine DOF has a stiffness degradation of 10%. In the first column of Table 56.1 the first three eigenfrequencies for the bending modes are shown with an accuracy of three correct decimal places. The solution of the eigenvalue problems determines the set of stiffness parameters as shown in the second column. A stiffness parameter equal to one refers to an undamaged structural element. The decrease of the second stiffness parameter points to a damage in this region. The value of 0.897 indicates a stiffness reduction of 10.3% which is quite close to the damage applied before. Furthermore, damage of the correct structural element is reflected since the stiffness parameters of elements one and three remain almost identical to one.

56.3 Validation of the Method

A validation of the method by means of a sensitivity analysis was carried out. Therewith the influence of several parameters of the method on the accuracy of the quantification was investigated. The parameters of the method investigated are, e.g. accuracy of eigenfrequencies, size of damaged area, stiffness reduction, location, number of damages, number of considered eigenfrequencies. It turned out that the crucial parameter, which essentially influences the method, is the accuracy of the eigenfrequencies. A residual analysis determines the best fit for the simulated data, which is achieved by a linear proportionality function. The confidence interval is unchanged over the area of validity as well as the limit of quantification. Thus detection and quantification have the same sensitivity.

56.4 Outlook

The next step of the validation of method is the confirmation of the results of the simulations on test structures. For this purpose scale test models will be established and tests will be carried out in the laboratory and under natural excitation. Before real application on wind turbines further tests on large structures with real or artificial damages are desirable.

References

1. Turek M, Ventura C (2005) Finite Element Model Updating of a Scale-Model Steel Frame Building. In: Proceedings of the 1st IOMAC
2. Cottin N (2001) Dynamic model updating–a multiparameter eigenvalue problem. Mechanical Systems and Signal Processing 15:649–665
3. Atkinson FV (1972) Multiparameter Eigenvalue Problems I – Matrices and Compact Operators. Academic, New York, London

57

Influence of the Type and Size of Wind Turbines on Anti-Icing Thermal Power Requirements for Blades

L. Battisti, R. Fedrizzi, S. Dal Savio and A. Giovannelli

Summary. Ice accretion on wind turbine components affects system safe operation and performance, with an annual power loss up to 20–50% at harsh sites. Therefore, special design requirements, as ice prevention systems, are necessary for wind turbines to operate in cold climates. The paper presents results for evaluating the anti-icing energy requirement for wind turbine blades as a function of wind turbine size and type. The analysis was carried out by means of the TREWICE$^{\circledR}$ code tool.

57.1 Introduction

Although relatively numerous aeronautical ice accretion–prediction codes have been developed , only a few dedicated codes exist for the analysis of ice accretion and the ice prevention systems design for wind turbines. Among them, the LEWICE 2.0 NASA code [1], adapted to predict anti-icing power requirements for wind turbines and the TURBICE code [2] that simulates ice accretion on wind turbine blades (VTT). In the few papers appeared these codes have been used to predict ice accretion of a single wind turbines blade section and to asses the main heat and mass flux contributions. When development and design of ice prevention systems is of concern, a lack of analysis and knowledge has been recognized by the authors of the paper on the following matters:

1. The correct identification of the parameters which mainly affect the performance and design characteristics of ice prevention systems
2. The effect of the selected anti-icing and de-icing strategy on turbine performance and loads
3. The effect of size and type of wind turbines on anti-ice thermal power requirements

The new TREWICE [3] code was used for analyzing and designing anti-icing systems for MW-size wind turbines with regard to the above three items.

57.2 Analysis of the Results

The investigation was focused on assessing the magnitude of the thermal anti-icing power, varying the size and type (1, 2 or 3 blades) of the wind turbine. In the latter case, the number of blades was changed keeping the rotor diameter and solidity constant. Under such an assumption, the velocity triangles change only because so does the Prandtl tip loss factor that explicitly depends on the number of blades. The rotors used were generated by means of a BEM code, which showed that for all the study cases considered, the angle of attack ranged between 2° and 12° from the middle to the tip. The environmental data reported in Table 57.1 were used to compute the anti-icing heat flux requirement for the chosen rotors. A temperature of 2°C was set on the blade's outer surface to create a running wet condition over the external heated surface.

57.3 Anti-Icing Power as a Function of the Machine Size

Four 3-bladed turbines with different rated power output were selected (see Table 57.2). The heated spanwise length along the blade radius L, and the width H of the heated zone are shown in Table 57.2.

Trials on ice accretion suggested keeping the same width of the heated zone for all wind turbines. In fact as the turbines size decreases, the chord length decreases as well, while the water collection increases. L corresponds to the profiled length of the blade for all cases. The anti-icing heat flux q_t, averaged on the heated blade's area, is reported in Fig. 57.1a. A larger heat flux has to be supplied to the bigger rotors since the heated area lies close to the leading edge where the heat exchange reaches a peak. Such a behaviour is emphasized for blades working at high AoA since higher convective heat exchange coefficient and water collection occur onto the blade's surface. The overall thermal power $Q_t = q_t * N_{\text{blades}} * A_{\text{heated}}$ (Fig. 57.1b) increases with the turbine size as a consequence of the increase of the area to be heated.

Table 57.1. Site variables used for the simulations

T_s (°C)	T_∞ (°C)	p_∞ (Pa)	R_h (−)	LWC (g m^{-3})	MVD

Table 57.2. Turbine data used for a comparison of the anti-icing thermal power with respect to machine size

	P_r (kW)	D (m)	Z (−)	N_r (rpm)	V_r (m s^{-1})	NACA airfoil	L (m)	H (m)
M1	726	40.76	3	33.21	12.5	44xx	16.1	0.32
M2	1,630	61.12	3	22.15	12.5	44xx	24.1	0.32
M3	3,668	91.68	3	14.76	12.5	44xx	36.2	0.32
M4	6,522	122.24	3	11.06	12.5	44xx	48.2	0.32

57 Influence of the Type and Size of Wind Turbines   307

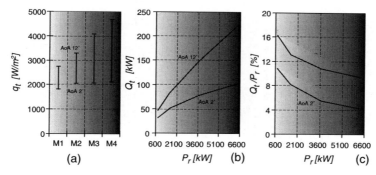

Fig. 57.1. Specific anti-icing power requirement (**a**), total anti-icing power requirement (**b**), and anti-icing-power to rated-power (**c**) between the two extreme angles of attack

Table 57.3. Turbine data used for a comparison of the anti-icing thermal power with respect to machine type

	P_r (kW)	D (m)	Z (–)	N_r (rpm)	V_r (m s^{-1})	NACA airfoil	L (m)	H (m)
M3	1,630	61.12	3	22.15	12.5	44xx	24.1	0.32
M5	1,571	61.12	2	22.15	12.5	44xx	24.1	0.32
M6	1,421	61.12	1	22.15	12.5	44xx	24.1	0.32

However, the ratio of anti-icing power to rated power output (Fig. 57.1c) is unfavourable for the smaller turbines, growing from 4 to 10% for the largest to 11–16% for the smallest one.

57.4 Anti-Icing Power as a Function of the Machine Type

Three wind turbines were considered having almost the same rated power and different number of blades (see Table 57.3).

Figure 57.2a shows that larger heat fluxes have to be supplied to the wider cord rotor (the single-bladed one) since the heated area lies close to the leading edge. However, the anti-icing power requirement Q_t (Fig. 57.2b), which depends on the number of blades, decreases considerably from the three-to the single-bladed turbine, in a manner that is nearly proportional to the number of blades. Figure 57.2c shows that the reduction of the number of blades reduces the ratio of rotor anti-icing thermal power to rated turbine power.

57.5 Conclusions

The presented analysis showed that three-bladed turbines with a rated power output between 1 and 6 MW, depending on environmental conditions and for

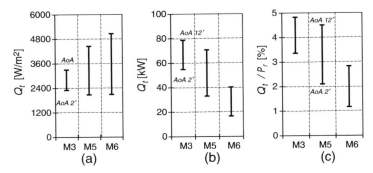

Fig. 57.2. Specific anti-icing power requirement (**a**), total anti-icing power requirement (**b**), and anti-icing-power to rated-power (**c**) for the two extreme angles of attack

a running wet condition, require an anti-icing power at the leading edge region increasing with the size and ranging between 10% and 15% of the turbine's rated power output. This ratio is lower for larger size turbines compared to small ones. The anti-icing power decreases more or less proportionally to the number of blades for turbines having the same solidity and size. The same conclusions, not shown here can be drawn for rotors of different solidity, although other physical mechanisms are involved. It is worth to mention that the above results concern the heat flux requirement and no considerations were developed at this stage on the adopted ice prevention system technology. The transfer of the results for their design must be carefully evaluated as the efficiency of the ice prevention system plays a basic role in defining the actual power requirement.

References

1. W.B. Wright, User Manual for the NASA Glenn Ice Accretion Code LEWICE version 2.2.2, CR-2002-211793, pp. 260–270, 363–366
2. L. Makkonen, T. Laakso, M. Marjaniemi, K.J. Finstad, Modeling and prevention of ice accretion on wind turbines, Wind Engineering, 2001, 25(1), 3–21
3. L. Battisti, R. Fedrizzi, M. Rialti, S. Dal Savio, A model for the design of hot-air based wind turbine ice prevention system, WREC05, 22–27 May 2005, Aberdeen, UK

58

High-cycle Fatigue of "Ultra-High Performance Concrete" and "Grouted Joints" for Offshore Wind Energy Turbines

L. Lohaus and S. Anders

58.1 Introduction

"Grouted Joints" are well known for offshore platforms in the oil and petroleum industry. Besides, they have already been used in Monopile-foundations and are being considered for Tripod-foundations in deeper water for offshore wind energy converters. In offshore wind energy applications, only ultra-high performance concrete (UHPC) has been used up to now. UHPC itself is known for its outstanding mechanical characteristics in static tests, but only little is known about the dynamic behaviour in terms of fatigue strength, damage development or the performance in "Grouted Joints."

58.2 Ultra-High Performance Concrete

UHPC is known for its distinct brittle failure and its outstanding compressive strength, but only little is known about its fatigue performance [3,4].

Experiments on different UHPC mixes [3], with compressive strengths ranging from 135 to 225 MPa, have indicated, that attention has to be paid to the fatigue strength, which seems to be lower for high-cycle fatigue compared to normal-strength concrete using normalized stresses, as illustrated in Fig. 58.1.

In contrast to the UHPC mix tested in Aalborg, heat treatment with different temperatures was used in Hannover and Kassel. In addition, the UHPC mix tested in Kassel contained 2.5 vol.% of 9 mm long steel fibres. The tests performed by the University of Kassel are displayed as a dashed line, because the number of tests was limited.

Deformation measurements and the development of Young's modulus during fatigue tests have indicated, that the distinct brittleness of UHPC can also be observed in fatigue loading. These results are in good agreement with deformation measurements on UHPC reported in [3].

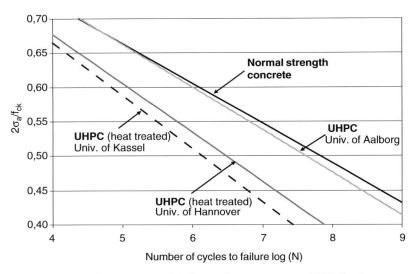

Fig. 58.1. Comparison of different fatigue tests on UHPC mixes

The advantage of calculating Young's modulus instead of the deformation development is that a degradation formula can be derived and implemented into Finite Element Programs. Different approaches for the development of Young's modulus for normal strength concrete without fibres have been proposed so far.

58.3 Ultra-High Performance Concrete in Grouted Joints

As stated for instance in the existing design rules for Grouted Joints [1, 2], specimens containing shear keys show an increasing ultimate load with an increasing shear key height. In the following static and fatigue tests on scaled-down Grouted Joints specimens will be discussed. These specimens were projected to comply as far as possible with the regulations published by Det Norske Veritas. Further conditions were commercially available steel profiles as well as the present testing devices.

In the static tests, specimens without shear keys and specimens with shear key heights of 0.3 and 1.25 mm were tested (Fig. 58.2 a).

It can be seen that the specimens containing shear keys show two nearly linear parts in the load–displacement curves. From the results obtained so far it seems as if the failure at the end of the first part is caused by a failure at the lowest compression strut between the shear keys of the pile and the sleeve. This point is referred to as first maximum load. With the load further increasing, the concrete on the compression side of the shear keys starts to crush and

58 High-cycle Fatigue   311

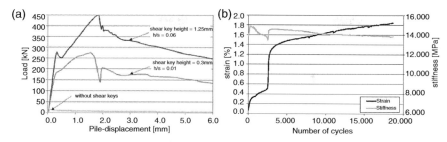

Fig. 58.2. (a) Static load–displacement curves for different Grouted Joints specimens (b) Strain and stiffness development of Grouted Joints in fatigue tests

when reaching the ultimate load the pile is gradually pushed through the grouted connection with considerable load-bearing capacity remaining. One aspect that should be noted, is the deformation at the ultimate load. Related to the grouted length of 90 mm, a pile displacement of about 2 mm equals a deformation of over 2%. In view of alternating tension and compression stresses in Grouted Joints for Tripod foundations, this might have detrimental effects on the fatigue strength as well as the deviations at the top of the tower. In this case, it might be sensible to apply the first maximum load for design purposes.

Comparative calculations according to the regulations of the American Petroleum Institute [1] and Det Norske Veritas [2] show that all of the tests are conservative compared to the American Petroleum Institute, but do not reach the characteristic values given by Det Norske Veritas.

Fatigue tests have supported that the load redistributes when reaching the first maximum or a comparable strain as displayed in Fig. 58.2b. At this point the strain increases from about 0.5% to 1.2% during a few load cycles. The following gradual increase corresponds to the second part of the static curve in Fig 58.2a. Regarding the stiffness development of Grouted Joints in fatigue tests the typical decrease in stiffness can be observed up to the first maximum load. At this point the stiffness recovers and nearly reaches its initial value, before gradually decreasing again.

58.4 Conclusions

In this paper the performance of UHPC in Grouted Joints subjected to static and fatigue loading is discussed. It is described (Fig 58.1), that the fatigue strength of UHPC in high-cycle fatigue seems to be lower compared to normal-strength concrete. Regarding UHPC in Grouted Joints nearly bilinear load–deflection curves for specimens with shear keys in uniaxial compression are shown, which represent two different load-bearing mechanisms and a very ductile failure. The redistribution of the load at the first maximum described for static tests can clearly be recognized in fatigue tests as well. Furthermore

312 L. Lohaus and S. Anders

the tests on downsized grouted connections indicate that the regulations of the American Petroleum Institute are conservative regarding the tested specimens whereas the regulations by Det Norske Veritas overestimate the first maximum load as well as the ultimate load. Further investigations are in progress to confirm these results.

Acknowledgements

The generous financial support of the Stiftung Industrieforschung, the provision of the steel tubes by Vallourec and Mannesmann Tubes and the fatigue results on UHPC specimens by Densit A/S is gratefully acknowledged.

References

1. American Petroleum Institute (2000): Recommended Practice for Planning, Designing and Constructing Fixed Offshore Platforms–Working Stress Design. API Washington
2. Det Norske Veritas (1998): Rules for Fixed Offshore Installations, Det Norske Veritas
3. E. Fehling, M. Schmidt et al. (2003): Entwicklung, Dauerhaftigkeit und Berechnung Ultra-Hochfester Betone, Forschungsbericht an die DFG, Projekt Nr. FE 497/1-1, Universität Kassel
4. L. Lohaus, S. Anders (2004): Effects of polymer- and fibre modifications on the ductility, fracture properties and micro-crack development of UHPC. International Symposium on UHPC, 13.-15.09.2004, Kassel

59

A Modular Concept for Integrated Modeling of Offshore WEC Applied to Wave-Structure Coupling

Kim Mittendorf, Martin Kohlmeier, Abderrahmane Habbar
and Werner Zielke

59.1 Introduction

For the development of robust and economic offshore wind energy converters, reliable design methods to perform integrated simulation of offshore wind turbines in time domain have to be improved or even developed.

In the offshore environment, we have to consider a coupled system, consisting of the support structure, the foundation, the aeroelastic system for the wind model and the hydrodynamic loads due to waves.

The achievement of an integral model suffers from the diversity of different processes and process interactions to be taken into account for the analysis of an offshore wind turbine and its associated subsystems. The models used by research teams and consulting engineers are normally heterogeneous. Therefore, a flexible structure of the integral model is the main target of current research. A well designed object-oriented and easily extendable set of models and interfaces have to be developed in order to fulfill future demands.

This paper gives an overview of the single modules which are needed for an integrated design and how they would interact. Moreover it describes the pursued strategy for the coupling procedures and the interface design. In the first step a model for fluid structure interaction is implemented.

59.2 Integrated Modeling

The numerical simulation of coupled processes (see Fig. 59.1) is essential for optimization purposes and for prediction of the breakeven performance of offshore WEC. A variety of commercial simulation tools for fluid dynamics, multibody dynamics or structural mechanics is available. Additional sophisticated tools for specific demands are being developed or self-improved. In order to serve several purposes an alternative use of different models is required in order to fulfill the a wide variety of modeling and simplification strategies.

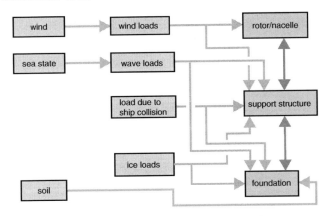

Fig. 59.1. Components of the integrated model approach: processes, loads, and subsystems

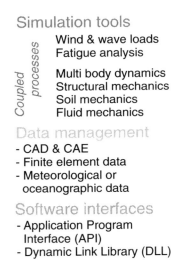

Fig. 59.2. Coupled processes and submodules of the IM

The set-up of a comprehensive model suffers from the multitude of heterogeneous modeling frameworks involved. The *integrated model* (IM), to be developed within the research activities of the *Center for Wind Energy Research* (ForWind), is supposed to reduce these shortcomings [2]. Especially, the data management and the preparation or automation of simulation sequences are important aspects (Fig. 59.2). Thus, the IM has to be developed in terms of a flexible compound of modular simulation models combined with control, data base, and interface units. These interfaces are one of the main targets as they will improve and faster interacting simulations. A modular concept will also enable independent code developments of several research teams.

59.2.1 Model Concept

The IM research work is aimed to get together the diversity of different processes and process interactions which have to be taken into account for the analysis of an offshore wind turbine and its associated subsystems. Therefore, a flexible structure of the IM is the main target of current research. A well designed object-oriented and easily extendable set of models and interfaces have to be developed in order to fulfill future demands. The current approach is depicted in Fig. 59.3.

59.2.2 Model Realization

The current status of development of the IM allows to define the geometric data, its discretization and the belonging material properties for the transformation to the load module (WeveLoads) followed by the specification of input files for a dynamic analysis using a finite element solver. Thus, the user is assisted in three ways: the geometric and material data transformation on the one hand and the transformation (global/local) of the loading data or evaluation of resulting values on the one hand. With increasing amount of integrated submodules the usefulness of having graphical modules available in the IM gets more apparent as well as the need of data format transition.

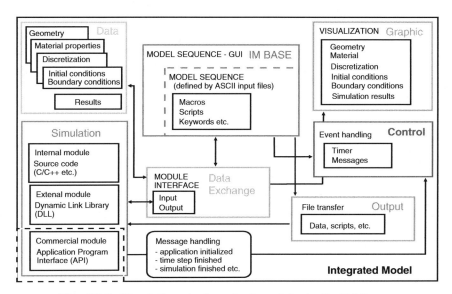

Fig. 59.3. Simulations tools and submodules in the IM frame

59.3 Modeling of Wave Loads on the Support Structure Offshore Wind Energy Turbines

During the design process of an offshore wind turbine the engineer has to consider the extreme load condition as well as the fatigue life. Following, the support structure for the WEC is modeled with a finite element approach using the program ANSYS. The hydrodynamic loads are calculated using different linear or nonlinear wave theories in combination with the Morison equation [1,4]. Morison equation is limited to hydrodynamically transparent structures and is a summation of drag and inertia forces

$$F = \frac{1}{2}\rho C_D D \int_{-h}^{\eta} u\,|u|\,\mathrm{d}z + C_M \rho \frac{\pi D^2}{4} \int_{-h}^{\eta} \frac{\partial u}{\partial t}\mathrm{d}z \tag{59.1}$$

where u is the water particle kinematic perpendicular to the structure axis, C_D and C_M hydrodynamic coefficients, D tube diameter and ρ fluid density.

59.3.1 Application to the Support Structure of an Offshore Wind Turbine

In the next step the structural behavior of a monopile (Fig. 59.4) loaded by irregular waves with and without directionality of the wave field is analyzed. A reduction of the loads and the structure response up to 20% can be observed, when the directionality is taken into account (Fig. 59.5) [3]. In further examinations the effect on the fatigue will be analyzed.

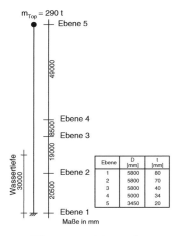

Fig. 59.4. Monopile

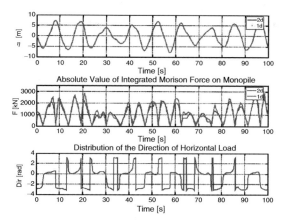

Fig. 59.5. 1D and 2D wave loads applied to monopile.

59.4 Future Demands

Future demands are not yet fully predictable, but a higher resolution of the used models also for substructures will yield in a demand for model adaptive strategies and new methods or solutions techniques.

The next milestone of the development will be the implementation of aeroelasticity model for the dynamical analysis of the structure due to combined wind and wave loads.

References

1. K. Mittendorf, B. Nguyen (2002): User Manual WaveLoads, Gigawind Report
2. M. Kohlmeier, ForWind (2004): Annual Report 2003/2004, TP IX
3. K. Mittendorf, H. Habbar, W. Zielke (2005): Zum Einfluss der Richtungsverteilung des Seegangs auf die Beanspruchung von OWEA, Gigawind Symposium Hannover
4. J.R. Morison, M.P. O'Brien, J.W. Johnson, S.A. Schaaf (1950): The Force exerted by Surface Waves on Piles, Pet. Trans., AIME, 189, pp. 149–154

60

Solutions of Details Regarding Fatigue and the Use of High-Strength Steels for Towers of Offshore Wind Energy Converters

J. Bergers, H. Huhn and R. Puthli

Summary. This article presents a new research project. The aim of this project is to evaluate the individual problems concerning stability and durability of towers of offshore wind energy converters and to develop new solutions. This includes the investigation of the influence of steel grade and wall thickness, the assessment of the remaining life and geometrical form of the connection. Another important aspect is the adjustment of the existing fatigue design concept to the special requirements of offshore wind energy converters.

60.1 Introduction

In the context of the exploiting alternative energies, many offshore wind parks are planned in the North Sea, Baltic Sea and Irish Sea. Aggravated conditions such as large water depths, increasing heights of towers and hubs, as well as other environmental conditions require new solutions concerning the stability and the durability of offshore wind energy converters (OWEC).

Therefore the Forschungsvereinigung Stahlanwendung e.V., sponsors a new research project [1]. The University of Karlsruhe as project leader is fortunate to have the involvement of competent partners. The Germanische Lloyd Windenergie GmbH as an accepted certification authority for the construction of OWEC takes care of corrosion problems. As a specialist in questions of durability the University of Braunschweig also participates in the project. The IMS Ingenieurgesellschaft mbH is a leading engineering company in the range of offshore constructions and is responsible for the specification of the structure and the numerical investigation. Furthermore, experts of wind energy and the steel industry also take part in the research project. The Research Centre for Steel, Timber and Masonry is mainly responsible for performing further finite-element analyses, determination of the most appropriate method of testing and the following fatigue investigation.

60.2 Fatigue Tests

It is generally assumed that fatigue strength decreases with an increasing wall thickness. To estimate the influence of wall thickness more exactly, however, there are different approaches depending on the national standards and the type of construction, showing a wide difference in the reduction factor (see Fig. 60.1).

When using high strength steel, smaller wall thicknesses usually suffice. Structures with many welds become more economical by thus also decreasing weld volume and post weld treatment for extending the remaining fatigue life. According to current standards, the choice of high strength steels with yield strengths up to 460 MPa is considered as mild steels and therefore no influence on the fatigue resistance can be taken into account. Fatigue resistance of higher steel grades is not considered in the standards.

To estimate the influence of yield strength, wall thickness and post weld treatment on the fatigue strength, various fatigue tests are carried out. It is known from former investigations at the University of Karlsruhe [7], that high strength steels up to a yield strength of 1,100 MPa do not show any disadvantage compared to mild steels regarding the fatigue strength. In several cases even better results for higher steel grades were achieved. For structures with low notch effects, high strength steel particularly exhibits better fatigue strength. One possibility to reduce the influence of the notch form is an effective post weld treatment, which is also investigated in the fatigue tests at present.

In Fig. 60.2 results from former investigations on crane specific details in Karlsruhe, investigations made in Italy [8] and the first results of the

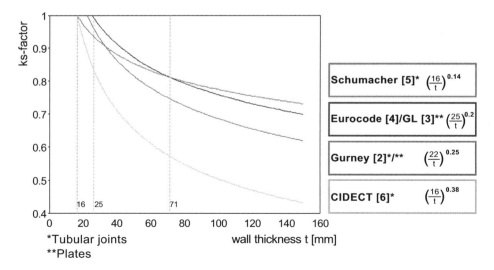

Fig. 60.1. Influence of wall thickness according to different standards

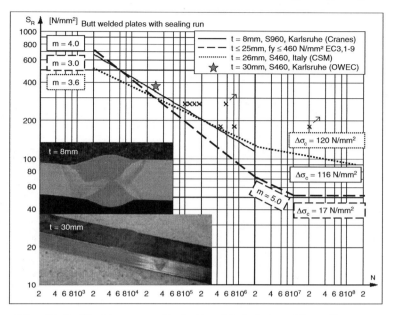

Fig. 60.2. Woehler-curves for butt welded plates with sealing run

present investigations for OWEC are compared. In addition, the corresponding Woehler-curve in EC3 part 1.9 is shown for comparison [4]. All Woehler-curves result from fatigue tests on butt welded plates with sealing runs. The diagram shows that the fatigue resistance for butt welded plates made of S960 is much higher. The result from the present test on 30 mm thick plates seems to fit into the Woehler-curve for 8 mm thick plates.

60.3 Finite-Element Analyses

The IMS Ingenieurgesellschaft mbH already analysed different structural forms with varying dimensions and came up with an innovative solution for the tower of OWEC via finite-element calculations.

In the first step of the analysis a global model was calculated for the ultimate and the fatigue limit state. This model includes all environmental loads acting on the foundation structure like wind and waves. The fatigue resistance of the joints is always the decisive design criteria for the lifetime of the structure. To carry out detail solutions regarding fatigue design further analyses were worked out via finite-element calculations. For this the forces and moments of different load cases were transferred from the global structure to the finite-element model. The main focus in this project lies on the construction of the upper central joint of the tripod, which is the most critical construction for fatigue design.

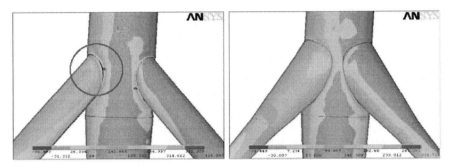

Fig. 60.3. Hot-spot stresses for cylindrical and conical connections at the upper central joint

One result of the detail analysis is shown in Fig. 60.3. For tubular connections hot-spot stresses at the transition between brace and chord cannot be avoided. The knowledge about stress distribution enables the localization of critical areas for partial damage calculation. The size of hot-spot stresses is depending on loading and tubular geometry. The hot-spot stresses can be calculated by use of stress concentration factors (SCF) or by finite-element methods.

To increase the fatigue resistance of the upper central joint the hot-spot stresses has to be reduced. One possible solution is shown in the right part of Fig. 60.3. Here a stress reduction is achieved by conical widening of the braces at transition.

Another possibility for a construction with high fatigue resistance is worked out together with crane constructors which have many experiences in fatigue design. An analogous connection to the upper central joint is solved in conveyor technique with a ring insert. In this construction the shear forces are carried off by internal stiffeners and not by plate bending, which reduce hot-spot stresses. Another effect is that the central shaft below the upper central joint is no longer needed, which also leads to an omission of the lower central joint and the lower braces.

This idea of an innovative construction without lower central joint is transferred to a tripod foundation structure for offshore wind turbines. At first a global and a detailed model of the upper central joint without ring insert was developed, see Fig. 60.4. The braces of this structure have the bearing behaviour of bending beams. Typical for tubular connections is that hot-spot stresses appear at tubular intersections.

Figure 60.5 shows the results of the stress calculation for the same load case as before for a tripod structure with ring insert. Here one can see that the hot-spot stresses are enormously reduced, which will lead to an efficient solution regarding fatigue design.

60 Solutions of Details Regarding Fatigue   323

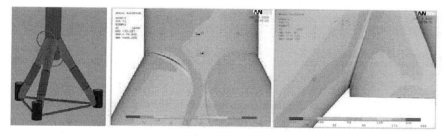

Fig. 60.4. Hot-spot stresses for tripod design without lower central joint

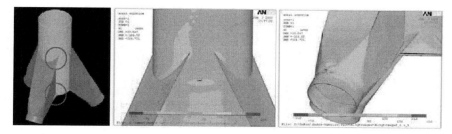

Fig. 60.5. Hot-spot stresses for tripod design without lower central joint and ring insert

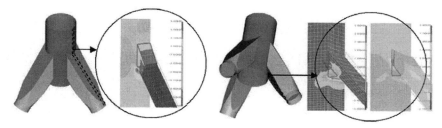

Fig. 60.6. Examples for FE-analyses on cut-outs of the tripod for finding a suitable test-setup

This specified innovative construction for the foundation structure of OWEC builds the basis for all further investigations in the new research project, which is now continued by material testing.

Structures with large wall thicknesses cause problems for the execution of the tests, so that simplified test specimens are investigated at first. In the second step, further tests on sections of the tripod joint are to be carried out. To determine the complex stress distribution in the tripod joint and to consider the surrounding conditions for the test setup, further comprehensive finite-element analyses are performed, as is illustrated in Fig. 60.3. In parallel, practical parameter studies are supported by extensive numerical work.

324   J. Bergers et al.

References

1. R. Puthli, M. Veselcic, S. Herion and H. Huhn: Detaillösungen bei Ermüdungsfragen und dem Einsatz hochfester Stähle bei Türmen von Offshore-Windenergieanlagen, Stahlbau, Heft 6, Ernst&Sohn Verlag, Berlin, Germany, June 2005
2. T.R. Gurney: The Some comments on fatigue gesign rules for offshore structures, 2nd International Symposium on Integrity of Offshore Structures, pp.219–234, Applied Science Publishers, Barking, Essex, England
3. Guideline for the Certification of Offshore Wind Turbines, Germanischer Lloyd, Edition 2005
4. prEN 1993-1-9, 2003-05: Eurocode 3: Design of steel structures – Part 1.9: Fatigue
5. A. Schumacher: Fatigue behaviour of welded circular hollow section joints in bridges, École Polytechnique Fédérale de Lausanne, doctoral thesis 2003
6. X.-L Zhoa et al.: Design guide for circular and rectangular hollow section joints under fatigue loading, TÜV-Verlag GmbH, 2000
7. Forschungsbericht P512: Beurteilung des Ermüdungsverhaltens von Krankonstruktionen bei Einsatz hoch- und ultrahochfester Stähle, Forschungsvereinigung Stahlanwendung e.V., 2005
8. E. Mecozzi et al.: Fatigue Behavior of Girth Welded Joints of High Grade Steel Risers, Proceedings of the Fourteenth International Offshore and Polar Engineering Conference. Toulon, France, 2004

61

On the Influence of Low-Level Jets on Energy Production and Loading of Wind Turbines

N. Cosack, S. Emeis and M. Kühn

61.1 Introduction

The variation of wind speed with height above ground is an important design parameter in wind energy engineering, as it influences power production and loading of the wind turbine. In a convective boundary layer we usually expect small vertical gradients of atmospheric variables like momentum or temperature. These unstable conditions are the normal case during daytime. At night, when the thermal production of turbulence vanishes and the mechanical production of turbulence is small (low to moderate wind speeds), the surface layer can decouple from the rest of the boundary layer. Such stable conditions lead to strong vertical gradients of atmospheric variables at night. Especially we observe very low wind speeds near the ground and higher wind speeds above the surface layer. This high wind speed band above the surface layer is called Low-Level-Jet (LLJ). In addition to the change in wind speed, the wind direction in the various layers can also differ significantly. In the following sections we evaluate the influence of LLJs that have been measured in the northern part of Germany on wind turbines by means of computer simulation.

61.2 Data and Methods

Since 1996, the Institute of Meteorology and Climatic Research of the Forschungszentrum Karlsruhe performed measurements of vertical wind profiles at various sites in Germany using Sodar. Details on the measurement campaigns and their results have been presented in [1, 2]. From the vast amount of data, two typical profiles of change in wind speed with height have been chosen for the described analysis. Although the change in wind direction with height has a significant impact on loading and power, only results from profiles with constant wind direction will be shown here.

The aero-elastic simulation tool Flex5 [3] has been used to estimate the power production and the loading for two turbine types: a 1.5 MW machine with 70 m rotor diameter and hub heights of 85, 100, and 130 m and a 5 MW machine with 128 m rotor diameter and hub heights of 100 m and 130 m. Both turbine types are typical pitch controlled variable speed wind turbines.

We applied a common approach for modeling the incoming wind: The deterministic part and the turbulent part of the wind are modeled separately, superposition of the two parts gives the full three dimensional turbulent wind field. The turbulent wind field has been modeled with a method described by Veers [4], using the Kaimal power spectral density functions as given in the IEC guidelines [5]. For both profiles, a standard deviation in longitudinal wind speed of $0.7\,\mathrm{m\,s^{-1}}$ has been estimated from measurements.

The center of the LLJ has been measured at 180 m, while the top of the surface layer is located at 50 m above the ground. In Profile A, the mean wind speed at a reference height of 85 m is $4\,\mathrm{m\,s^{-1}}$ with an additional 25% increase at the center of the LLJ, compared to what one would estimate solely from terrain and stratification. In Profile B, these parameters are $7\,\mathrm{m\,s^{-1}}$ and 6%.

61.3 Results

Two scenarios have been examined, S1 and S2. In S1, the results of simulations with the profiles A and B have been compared to results when the vertical wind profile has been modeled only according to the logarithmic law and stratification. This corresponds to the error that we would make, if we had to estimate the turbines performance by just knowing the wind speed at hub height and some basic environmental parameters, but would not be aware of the LLJ. The investigated sensors are power production, the standard deviation of the blades flapwise bending moment and the mean tilt moment at the nacelle center.

Figure 61.1 shows the relative deviations due to the LLJ. The influence is visible in all sensors, although the difference in power and blade bending moment stays below 5%. As expected, profile A gives higher differences compared to profile B. This is due to the stronger increase in wind speed from surface layer to the center of the LLJ. Also, the results for the 5 MW turbine are generally higher than those for the 1.5 W turbine, due to the bigger rotor plane. The influence of the LLJ is highest in the case of the 5 MW turbine with a hub height of 130 m and profile A.

To make things worse, we assumed in S2 that the given wind speed has not been measured at hub height as before, but at some point below. This is often the case, when the wind resource at a site is evaluated during the planning of a wind park or when too small measurement masts are used in a wind park for performance evaluation. In comparison to S1, the difference between the situations with and without LLJ is not only the wind profile. The LLJ now results also in an increased mean wind speed at hub height and, as the standard deviation is kept constant, reduced turbulence intensity. At reference height of 85 m the wind speed for both situations is equal.

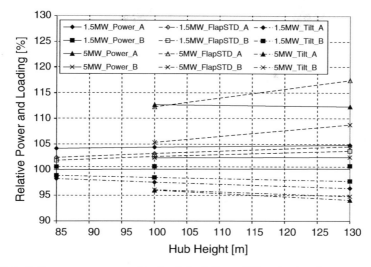

Fig. 61.1. Influence on power, standard deviation of flapwise bending moment and mean tilt moment for Scenario S1

The effect of this scenario is shown in Fig. 61.2. It can be seen, that the presence of a LLJ leads to much higher differences then in the previously examined situation. The difference in power output in the case of the 5MW turbine with 130 m hub height is almost 30%, while the difference in the blade bending moment's standard deviation is up to 25%. The general tendencies that have been observed before remain unchanged: the relative differences are higher for the 5 MW turbine and for profile A.

61.4 Conclusions

The main conclusions that can be drawn from the analysis are as follows:

- LLJs have clearly an impact on the turbines performance. They lead to an increased power output and standard deviation of the blades flapwise bending moments
- The influence of LLJs increases with turbine size
- The error in estimating the turbines performance can be quite high, up to 30% in the investigated cases, when the mean wind speed at hub height needs to be calculated from some height below

Regarding the influence of LLJ on life time calculations, it has to be considered that life-time fatigue loading and annual energy yield depend directly on the site-specific frequency of occurrence and severity of such events. Therefore their impact is hard to estimate but will probably be small. This is mainly due to the fact that at least at the measurement sites, these situations did

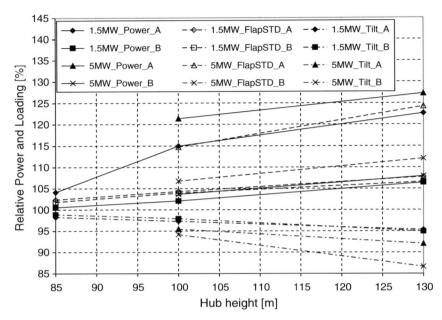

Fig. 61.2. Influence on power, standard deviation of flapwise bending moment and mean tilt moment for Scenario S2

only occur at low to moderate wind speeds. Furthermore the atmospheric conditions need to be stable, which results in relatively low turbulence intensities, compared to the unstable "normal" conditions. Finally, the overall time share at which these events take place is probably small compared to the turbines life time.

References

1. Emeis S (2001) Vertical variation of frequency distributions of wind speed in and above the surface layer observed by sodar, Meteorologische Zeitschrift, vol. 10, no. 2:141–149
2. Emeis S (2004) Vertical wind profiles over an urban area, Meteorologische Zeitschrift, vol. 13, no. 5:353–259
3. Oye S (1991) Simulation of Loads on a Wind Turbine including Turbulence, Proceedings of IEA-Meeting "wind characteristics of relevance for wind turbine design," Stockholm
4. Veers P S (1988) Three-Dimensional Wind Simulation, Sandia Report, SAND88–0152
5. IEC 61400-1(1998) Wind Turbine Generator Systems Part1: Safety requirements

62

Reliability of Wind Turbines
Experiences of 15 years with 1,500 WTs

Berthold Hahn, Michael Durstewitz and Kurt Rohrig

62.1 Introduction

With the rapid expansion of wind energy use in Germany over the past 15 years, extensive developments in wind turbine (WT) technology have taken place. The new technology has achieved such a level of quality, that WTs obtain a technical availability of 98%.

This means that an average WT will be inactive for around 1 week per year for repairs and maintenance. Considering that the WTs operate over years without operating personnel, this average downtime seems short. The paper gives some figures about reliability of wind turbines, failures and downtimes for wind turbines and components.

62.2 Data Basis

In the framework of the "250 MW Wind" Programme, ISET is monitoring over 1,500 WTs in operation. Over a period of 15 years now, WTs with a variety of different technical conceptions and installed in different regions in Germany have been included in the programme. So, from these turbines, the experiences of up to 15 operating years are readily available. On average, the participating turbines have completed 10 years of operation.

Operators of the supported WTs regularly report to ISET concerning energy yields, maintenance and repairs, and operating costs. In form sheets for maintenance and repair, the operators report on the downtimes caused by malfunctions, the damaged components and – as far as possible – the causes and obvious effects on turbines and operation.

Most of these supported WTs have a rated power below 1 MW. Thus, in the recent years operators of megawatt WTs where asked quite successfully to contribute to the programme on a voluntary basis. So, experiences of the recent models can be included into the evaluations as well.

Up to now, over 60,000 reports on maintenance and repair have been submitted to ISET. Standardized evaluations are published in the "Wind Energy Report" [1], which is updated annually.

62.3 Break Down of Wind Turbines

Usually, WTs are designed to operate for a period of 20 years. But, no final statement can be made yet concerning the actual life expectancy of modern WTs as, until now, no operational experience of such period is available. Changes in reliability with increasing operational age can, however, provide indications of the expected lifetime and the amount of upkeep required. Reliability can be expressed by the number of failures per unit of time, i.e. "Failure Rate." In the following, the failure rates of WTs depending on their operational age will be depicted (Fig. 62.1).

It is clear that the failure rates of the WTs now installed, have almost continually declined in the first operational years. This is true for the older turbines under 500 kW and for the 500/600 kW class. However, the group of megawatt WTs show a significantly higher failure rate, which also declines by increasing age. But, including now more and more megawatt WT models of the newest generation, the failure rate in the first year of operation is being reduced.

The principal development of failure rates is well known in other technical areas. "Early failures" often mark the beginning of operation. This phase is generally followed by a longer period of "random failures," before the failure rate through wear and damage accumulation ("wear-out failures") increases with operational age.

The total life period and the individual phases are naturally distinct for different technical systems. For WTs, hardly any experience is available in this

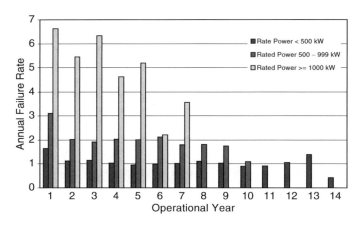

Fig. 62.1. Frequency of "failure rate" with increasing operational age

respect. Based on the above evaluations, however, for the WTs under 500 kW it can be expected that the failure rate due to "wear-out failures" does not increase before the 15th year of operation.

62.4 Malfunctions of Components

The reported downtimes are caused by both regular maintenance and unforeseen malfunctions. The following evaluations refer only to the latter, which concerned half mechanical and half electrical components (Fig. 62.2).

Besides failure rates, the downtimes of the machines after a failure are an important value to describe the reliability of a machine.

The duration of downtimes, caused by malfunctions, are dependent on necessary repair work, on the availability of replacement parts and on the personnel capacity of service teams. In the past, repairs to generator [2], drive train, hub, gearbox and blades have often caused standstill periods of several weeks.

Taking into account all the reported repair measures now available, the average failure rate and the average downtime per component can be given (Fig. 62.3). It gets clear, that the downtimes declined in the past 5–10 years. So, the high number of failures of some components is now balanced out to

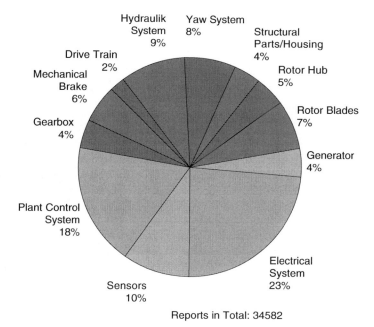

Fig. 62.2. Share of main components of total number of failures

332 B. Hahn et al.

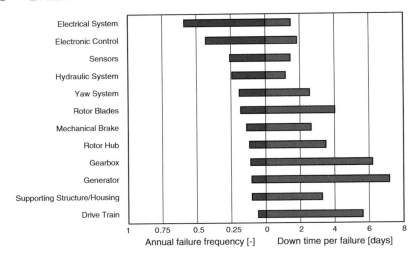

Fig. 62.3. Failure Frequency and downtimes of components

a certain extent by short standstill periods. But still, damages of generators, gear boxes, and drive trains are of high relevance due to long downtimes of about 1 week as an average.

62.5 Conclusion

Wind turbines achieve an excellent technical availability of about 98% on average, although they have to face a high number of malfunctions. It can be assumed that these good availability figures can only be achieved by a high number of service teams who respond to turbine failures within short time. In order to further improve the reliability of WTs, the designers have to better the electric and electronic components. This is particularly true and absolutely necessary in the case of new and large turbines.

References

1. C Ensslin, M Durstewitz, B Hahn, B Lange, K Rohrig (2005) German Wind Energy Report 2005. ISET, Kassel
2. M Durstewitz, R Wengler (1998) Analyses of Generator Failure of Wind Turbines in Germanys '250 MW Wind' Programme, Study. ISET, Kassel

Printed by Publishers' Graphics LLC
BT20130307.19.20.104